guilty knowledge

What the US Government Knows about the Vulnerability of the Electric Grid, But Refuses to Fix

CENTER FOR SECURITY POLICY PRESS

ARCHIVAL SERIES

Guilty knowledge: What the US Government Knows about the Vulnerability of the Electric Grid, But Refuses to Fix is published in the United States by the Center for Security Policy Press, a division of the Center for Security Policy.

THE CENTER FOR SECURITY POLICY
1901 Pennsylvania Avenue, Suite 201 Washington, DC 20006
Phone: (202) 835-9077 | Email: info@securefreedom.org
For more information, please see securefreedom.org

Book design by David Reaboi and Adam Savit

Contents

Foreword

On January 21, 2014, Fox News aired a segment describing the vulnerability of the U.S. bulk power distribution system, popularly known as the electric "grid." The report described various dangers that could cause the grid to fail, possibly catastrophically. These range from physical and cyber attacks on its subsystems to space weather and a high-altitude nuclear detonation unleashing intense electromagnetic pulses (EMP) that could afflict the grid across vast areas.

Fox News solicited a comment from the Department of Defense about these threats and their potential to imperil the very existence of the United States—and a large percentage of its present population. This was the Pentagon's response: "The Department is unaware of any increase in the threat of a deliberate destructive use of an EMP device. Further, any reporting to the contrary by those without access to current threat assessments is both reckless and irresponsible."

At the very best, this statement suggests that the Defense Department is ignorant of a yawning danger to the civilian critical infrastructure—upon which the military also heavily relies. At worst, it is actively and purposefully misleading the American people who will die by the tens of millions when one or the other of these threats eventuates.

In fact, a blue-ribbon commission convened by the Congress to examine the EMP threat concluded that, if the power went out and stayed off for more than a year in large parts of the United States—a prospect it found was plausible—as many as nine-out-of-ten Americans would perish.

Even if it actually *were* the case that EMP threats are not intensifying—something that is highly debatable in light of evidence in the public domain about the North Korean and Iranian nuclear weapons, ballistic missile and satellite programs—one thing is clear: U.S. civil society has been for many years so dangerously vulnerable to the take-down of the nation's electric grid as to *invite* enemies to try to exploit our vulnerability.

Moreover, even if no enemies acted on this opportunity to bring about, in the oft-stated words of then-Iranian president Mahmoud Ahmadinejad, "a world without America," there is another menace that is certain to do that, somewhat later if not quite soon: a massive geo-magnetic disturbance (GMD). Such a powerful GMD would distort the earth's magnetosphere, unleashing what are known as E3 long-duration electromagnetic pulses that would, all other things being equal, be con-ducted by power lines into the backbone of the grid: the nation's high-voltage trans-formers, seriously damaging if not destroying them.

In other words, the vulnerability of America's grid does not have to become any more severe to pose a mortal danger. To pretend otherwise—and to encourage the public to believe a false narrative—is what is truly "reckless and irresponsible."

That is especially so since the Department of Defense and the rest of the United States government have ample evidence of this peril. No fewer than eleven studies conducted by or for federal agencies in the past decade have provided an extraordinary consensus: **The nation's bulk power distribution system can be disrupted or destroyed over large areas due to various man-caused and naturally occurring phenomena.**

Should one or more of these types of events occur, there could be prolonged blackouts afflicting much of the country. That would deny millions of people, perhaps for years, the services they depend upon from more than a dozen critical infrastructures—all of which require electricity to operate for more than a few days. Without such services for that length of time, there will be massive loss of life and societal breakdown.

There is no basis for official professions of ignorance about the very much *present* danger posed by EMP, space weather, cyber warfare or direct physical attacks on the grid, or the dire effect these attacks could have on much of our population.

Think of it as "guilty knowledge": Knowing that this existential threat exists, one has a duty to ensure that the steps required to remediate it are taken.

In the interest of ensuring that the rest of us have ready access to this knowledge, the Center for Security Policy has compiled in one short reference book the executive summaries of these eleven studies. The full text of each may be viewed at the web site of the EMP Coalition (StopEMP.org), a group sponsored by the Center. Under the leadership of its Honorary Co-Chairmen, former House Speaker Newt Gingrich and former Clinton Director of Central Intelligence R. James Woolsey, the Coalition is working to raise public awareness of the electric grid's myriad vulnerabilities and to achieve the needed corrective action.

Our hope is that this compendium will make clear the abundant evidence distilled from authoritative sources that confirms America has a problem: We are at risk of unprecedented catastrophe from long-duration disruption of the electric grid—unless we take practical, near-term and relatively low-cost steps to prevent it. (More information about these steps is available at StopEMP.org.)

Equipped with this guilty knowledge, we hope you will recognize and act upon the duty to yourself, your family, your community and your country to ensure that the steps needed to make our grid resilient are taken, before it is too late.

Frank J. Gaffney, Jr.
President and CEO
Center for Security Policy

Report of the Commission to Assess the Threat to the United States from Electromagnetic Pulse (EMP) Attack

Volume 1: Executive Report

Commission Members: Dr. John S. Foster, Jr. | Mr. Earl Gjelde | Dr. William R. Graham (Chairman) | Dr. Robert J. Hermann | Mr. Henry (Hank) M. Kluepfel | Gen Richard L. Lawson, USAF (Ret.) | Dr. Gordon K. Soper | Dr. Lowell L. Wood, Jr. | Dr. Joan B. Woodard

2004

Highlights:

"Several potential adversaries have or can acquire the capability to attack the United States with a high-altitude nuclear weapon-generated electromagnetic pulse (EMP). A determined adversary can achieve an EMP attack capability without having a high level of sophistication."

"The electromagnetic fields produced by weapons designed and deployed with the intent to produce EMP have a high likelihood of damaging electrical power systems, electronics, and information systems upon which American society depends. Their effects on dependent systems and infrastructures could be sufficient to qualify as catastrophic to the Nation."

Found online at http://www.empcommission.org/docs/empc_exec_rpt.pdf

ABSTRACT

Several potential adversaries have or can acquire the capability to attack the United States with a high-altitude nuclear weapon-generated electromagnetic pulse (EMP). A determined adversary can achieve an EMP attack capability without having a high level of sophistication.

EMP is one of a small number of threats that can hold our society at risk of catastrophic consequences. EMP will cover the wide geographic region within line of sight to the nuclear weapon. It has the capability to produce significant damage to critical infrastructures and thus to the very fabric of US society, as well as to the ability of the United States and Western nations to project influence and military power.

The common element that can produce such an impact from EMP is primarily electronics, so pervasive in all aspects of our society and military, coupled through critical infrastructures. Our vulnerability is increasing daily as our use of and dependence on electronics continues to grow. The impact of EMP is asymmetric in relation to potential protagonists who are not as dependent on modern electronics.

The current vulnerability of our critical infrastructures can both invite and reward attack if not corrected. Correction is feasible and well within the Nation's means and resources to accomplish.

EMP IS CAPABLE OF CAUSING CATASTROPHE FOR THE NATION

The high-altitude nuclear weapon-generated electromagnetic pulse (EMP) is one of a small number of threats that has the potential to hold our society seriously at risk and might result in defeat of our military forces.

Briefly, a single nuclear weapon exploded at high altitude above the United States will interact with the Earth's atmosphere, ionosphere, and magnetic field to produce an electromagnetic pulse (EMP) radiating down to the Earth and additionally create electrical currents in the Earth. EMP effects are both direct and indirect. The former are due to electromagnetic "shocking" of electronics and stressing of electrical systems, and the latter arise from the damage that "shocked"—upset, damaged, and destroyed—electronics controls then inflict on the systems in which they are embedded. The indirect effects can be even more severe than the direct effects.

The electromagnetic fields produced by weapons designed and deployed with the intent to produce EMP have a high likelihood of damaging electrical power systems, electronics, and information systems upon which American society depends. Their effects on dependent systems and infrastructures could be sufficient to qualify as catastrophic to the Nation.

Depending on the specific characteristics of the attacks, unprecedented cascading failures of our major infrastructures could result. In that event, a regional or national recovery would be long and difficult and would seriously degrade the safety and overall viability of our Nation. The primary avenues for catastrophic damage to the Nation are through our electric power infrastructure and thence into our telecommunications, energy, and other infrastructures. These, in turn, can seriously impact other important aspects of our Nation's life, including the financial system; means of getting food, water, and medical care to the citizenry; trade; and production of goods and services. The recovery of any one of the key national infrastructures is dependent on the recovery of others. The longer the outage, the more problematic and uncertain the recovery will be. It is possible for the functional outages to become mutually reinforcing until at some point the degradation of infrastructure could have irreversible effects on the country's ability to support its population.

EMP effects from nuclear bursts are not new threats to our nation. The Soviet Union in the past and Russia and other nations today are potentially capable of creating these effects. Historically, this application of nuclear weaponry was mixed with a much larger population of nuclear devices that were the primary source of destruction, and thus EMP as a weapons effect was not the primary focus. Throughout the Cold War, the United States did not try to protect its civilian infrastructure against either the physical or EMP impact of nuclear weapons, and instead depended on deterrence for its safety.

What is different now is that some potential sources of EMP threats are difficult to deter—they can be terrorist groups that have no state identity, have only one or a few weapons, and are motivated to attack the US without regard for their own safety. Rogue states, such as North Korea and Iran, may also be developing the capability to pose an EMP threat to the United States, and may also be unpredictable and difficult to deter.

Certain types of relatively low-yield nuclear weapons can be employed to generate potentially catastrophic EMP effects over wide geographic areas, and designs for variants of such weapons may have been illicitly trafficked for a quarter-century.

China and Russia have considered limited nuclear attack options that, unlike their Cold War plans, employ EMP as the primary or sole means of attack. Indeed, as recently as May 1999, during the NATO bombing of the former Yugoslavia, high-ranking members of the Russian Duma, meeting with a US congressional delegation to discuss the Balkans conflict, raised the specter of a Russian EMP attack that would paralyze the United States.

Another key difference from the past is that the US has developed more than most other nations as a modern society heavily dependent on electronics, telecommunications, energy, information networks, and a rich set of financial and transportation systems that leverage modern technology. This asymmetry is a source of substantial economic, industrial, and societal advantages, but it creates vulnerabilities and critical interdependencies that are potentially disastrous to the United States. Therefore, terrorists or state actors that possess relatively unsophisticated missiles armed with nuclear weapons may well calculate that, instead of destroying a city or military base, they may obtain the greatest political-military utility from one or a few such weapons by using them—or threatening their use—in an EMP attack.

The current vulnerability of US critical infrastructures can both invite and reward attack if not corrected; however, correction is feasible and well within the Nation's means and resources to accomplish.

Report of the Commission to Assess the Threat to the United States from Electromagnetic Pulse (EMP) Attack

Critical National Infrastructures

Commission Members: Dr. John S. Foster, Jr. | Mr. Earl Gjelde | Dr. William R. Graham (Chairman) | Dr. Robert J. Hermann | Mr. Henry (Hank) M. Kluepfel | Gen Richard L. Lawson, USAF (Ret.) | Dr. Gordon K. Soper | Dr. Lowell L. Wood, Jr. | Dr. Joan B. Woodard

2008

Highlights:

"When a nuclear explosion occurs at high altitude, the EMP signal it produces will cover the wide geographic region within the line of sight of the detonation. This broad band, high amplitude EMP, when coupled into sensitive electronics, has the capability to produce widespread and long lasting disruption and damage to the critical infrastructures that underpin the fabric of U.S. society."

"Because of the ubiquitous dependence of U.S. society on the electrical power system, its vulnerability to an EMP attack, coupled with the EMP's particular damage mechanisms, creates the possibility of long-term, catastrophic consequences."

Found online at http://www.empcommission.org/docs/A2473-EMP_Commission-7MB.pdf

PREFACE

The physical and social fabric of the United States is sustained by a system of systems; a complex and dynamic network of interlocking and interdependent infrastructures ("critical national infrastructures") whose harmonious functioning enables the myriad actions, transactions, and information flow that undergird the orderly conduct of civil society in this country. The vulnerability of these infrastructures to threats—deliberate, accidental, and acts of nature—is the focus of greatly heightened concern in the current era, a process accelerated by the events of 9/11 and recent hurricanes, including Katrina and Rita.

This report presents the results of the Commission's assessment of the effects of a high altitude electromagnetic pulse (EMP) attack on our critical national infrastructures and provides recommendations for their mitigation. The assessment is informed by analytic and test activities executed under Commission sponsorship, which are discussed in this volume. An earlier executive report, *Report of the Commission to Assess the Threat to the United States from Electromagnetic Pulse (EMP) — Volume 1: Executive Report* (2004), provided an overview of the subject.

The electromagnetic pulse generated by a high altitude nuclear explosion is one of a small number of threats that can hold our society at risk of catastrophic consequences. The increasingly pervasive use of electronics of all forms represents the greatest source of vulnerability to attack by EMP. Electronics are used to control, communicate, compute, store, manage, and implement nearly every aspect of United States (U.S.) civilian systems. When a nuclear explosion occurs at high altitude, the EMP signal it produces will cover the wide geographic region within the line of sight of the detonation.[1] This broad band, high amplitude EMP, when coupled into sensitive electronics, has the capability to produce widespread and long lasting disruption and damage to the critical infrastructures that underpin the fabric of U.S. society.

Because of the ubiquitous dependence of U. S. society on the electrical power system, its vulnerability to an EMP attack, coupled with the EMP's particular damage mechanisms, creates the possibility of long-term, catastrophic consequences. The implicit invitation to take advantage of this vulnerability, when coupled with increasing proliferation of nuclear weapons and their delivery systems, is a serious concern. A single EMP attack may seriously degrade or shut down a large part of the electric power grid in the geographic area of EMP exposure effectively instantaneously. There is also a possibility of functional collapse of grids beyond the exposed area, as electrical effects propagate from one region to another.

The time required for full recovery of service would depend on both the disruption and damage to the electrical power infrastructure and to other national infra-

[1] For example, a nuclear explosion at an altitude of 100 kilometers would expose 4 million square kilometers, about 1.5 million square miles, of Earth surface beneath the burst to a range of EMP field intensities.

structures. Larger affected areas and stronger EMP field strengths will prolong the time to recover. Some critical electrical power infrastructure components are no longer manufactured in the United States, and their acquisition ordinarily requires up to a year of lead-time in routine circumstances. Damage to or loss of these components could leave significant parts of the electrical infrastructure out of service for periods measured in months to a year or more. There is a point in time at which the shortage or exhaustion of sustaining backup systems, including emergency power supplies, batteries, standby fuel supplies, communications, and manpower resources that can be mobilized, coordinated, and dispatched, together lead to a continuing degradation of critical infrastructures for a prolonged period of time.

Electrical power is necessary to support other critical infrastructures, including supply and distribution of water, food, fuel, communications, transport, financial transactions, emergency services, government services, and all other infrastructures supporting the national economy and welfare. Should significant parts of the electrical power infrastructure be lost for any substantial period of time, the Commission believes that the consequences are likely to be catastrophic, and many people may ultimately die for lack of the basic elements necessary to sustain life in dense urban and suburban communities. In fact, the Commission is deeply concerned that such impacts are likely in the event of an EMP attack unless practical steps are taken to provide protection for critical elements of the electric system and for rapid restoration of electric power, particularly to essential services. The recovery plans for the individual infrastructures currently in place essentially assume, at worst, limited upsets to the other infrastructures that are important to their operation. Such plans may be of little or no value in the wake of an EMP attack because of its long-duration effects on all infrastructures that rely on electricity or electronics.

The ability to recover from this situation is an area of great concern. The use of automated control systems has allowed many companies and agencies to operate effectively with small work forces. Thus, while manual control of some systems may be possible, the number of people knowledgeable enough to support manual operations is limited. Repair of physical damage is also constrained by a small work force. Many maintenance crews are sized to perform routine and preventive maintenance of high-reliability equipment. When repair or replacement is required that exceeds routine levels, arrangements are typically in place to augment crews from outside the affected area. However, due to the simultaneous, far-reaching effects from EMP, the anticipated augmenters likely will be occupied in their own areas. Thus, repairs normally requiring weeks of effort may require a much longer time than planned.

The consequences of an EMP event should be prepared for and protected against to the extent it is reasonably possible. Cold War-style deterrence through mutual assured destruction is not likely to be an effective threat against potential protagonists that are either failing states or transnational groups. Therefore, making preparations to manage the effects of an EMP attack, including understanding what has happened, maintaining situational awareness, having plans in place to recover, challenging and exercising those plans, and reducing vulnerabilities, is critical to reducing the consequences, and thus probability, of attack. The appropriate national-level approach should balance prevention, protection, and recovery.

The Commission requested and received information from a number of Federal agencies and National Laboratories. We received information from the North American Electric Reliability Corporation, the President's National Security Telecommunications Advisory Committee, the National Communications System (since absorbed by the Department of Homeland Security), the Federal Reserve Board, and the Department of Homeland Security. Early in this review it became apparent that only limited EMP vulnerability testing had been accomplished for modern electronic systems and components. To partially remedy this deficit, the Commission sponsored illustrative testing of current systems and infrastructure components. The Commission's view is that the Federal Government does not today have sufficiently robust capabilities for reliably assessing and managing EMP threats.

The United States faces a long-term challenge to maintain technical competence for understanding and managing the effects of nuclear weapons, including EMP. The Department of Energy and the National Nuclear Security Administration have developed and implemented an extensive Nuclear Weapons Stockpile Stewardship Program over the last decade. However, no comparable effort was initiated to understand the effects that nuclear weapons produce on modern systems. The Commission reviewed current national capabilities to understand and to manage the effects of EMP and concluded that the Country is rapidly losing the technical competence in this area that it needs in the Government, National Laboratories, and Industrial Community.

An EMP attack on the national civilian infrastructures is a serious problem, but one that can be managed by coordinated and focused efforts between industry and government. It is the view of the Commission that managing the adverse impacts of EMP is feasible in terms of time and resources. A serious national commitment to address the threat of an EMP attack can develop a national posture that would significantly reduce the payoff for such an attack and allow the United States to recover in a timely manner if such an attack were to occur.

Severe Space Weather Events
Understanding Societal and Economic Impacts

NATIONAL RESEARCH COUNCIL OF THE NATIONAL ACADEMIES

Committee on the Societal and Economic Impacts of Severe Space Weather Events | Space Studies Board | Division on Engineering and Physical Sciences

2008

Highlights:

"The Carrington event is by several measures the most severe space weather event on record. It produced several days of spectacular auroral displays, even at unusually low latitudes, and significantly disrupted telegraph services around the world."

"While the socioeconomic impacts of a future Carrington event are difficult to predict, it is not unreasonable to assume that an event of such magnitude would lead to much deeper and more widespread socioeconomic disruptions than occurred in 1859, when modern electricity-based technology was still in its infancy."

Found online at http://www.nap.edu/openbook.php?record_id=12507

SUMMARY

SOCIETAL CONTEXT

Modern society depends heavily on a variety of technologies that are susceptible to the extremes of space weather—severe disturbances of the upper atmosphere and of the near-Earth space environment that are driven by the magnetic activity of the Sun. Strong auroral currents can disrupt and damage modern electric power grids and may contribute to the corrosion of oil and gas pipelines. Magnetic storm-driven ionospheric density disturbances interfere with high-frequency (HF) radio communications and navigation signals from Global Positioning System.

The effects of space weather on modern technological systems are well documented in both the technical literature and popular accounts. Most often cited perhaps is the collapse within 90 seconds of northeastern Canada's Hydro-Quebec power grid during the great geomagnetic storm of March 1989, which left millions of people without electricity for up to 9 hours. This event exemplifies the dramatic impact that extreme space weather can have on a technology upon which modern society in all of its manifold and interconnected activities and functions critically depends.

Nearly two decades have passed since the March 1989 event. During that time, awareness of the risks of extreme space weather has increased among the affected industries, mitigation strategies have been developed, new sources of data have become available (e.g., the upstream solar wind measurements from the Advanced Composition Explorer), new models of the space environment have been created, and a national space weather infrastructure has evolved to provide data, alerts, and forecasts to an increasing number of users.

Now, 20 years later and approaching a new interval of increased solar activity, how well equipped are we to manage the effects of space weather? Have recent technological developments made our critical technologies more or less vulnerable? How well do we understand the broader societal and economic impacts of extreme space weather events? Are our institutions prepared to cope with the effects of a "space weather Katrina," a rare, but according to the historical record, not inconceivable eventuality? On May 22 and 23, 2008, a workshop held in Washington, D.C., under the auspices of the National Research Council brought together representatives of industry, the federal government, and the social science community to explore these and related questions. This report was prepared by members of the ad hoc committee that organized the workshop, and it summarizes the key themes, ideas, and insights that emerged during the days of presentations and discussions.

THE IMPACT OF SPACE WEATHER

Modern technological society is characterized by a complex interweave of dependencies and interdependencies among its critical infrastructures. A complete picture of the socioeconomic impact of severe space weather must include both direct, industry-specific effects (such as power outages and spacecraft anomalies) and the collateral effects of space-weather-driven technology failures on dependent infrastructures and services.

Industry-specific Space Weather Impacts

The main industries whose operations can be adversely affected by extreme space weather are the electric power, spacecraft, aviation, and GPS-based positioning industries. The March 1989 blackout in Quebec and the forced outages of electric power equipment in the northeastern United States remain the classic example of the impact of a severe space weather event on the electric power industry. Several examples of the impact of space weather on the other industries are cited in the report:

* The outage in January 1994 of two Canadian telecommunications satellites during a period of enhanced energetic electron fluxes at geosynchronous orbit, disrupting communications services nationwide. The first satellite recovered in a few hours; recovery of the second satellite took 6 months and cost $50 million to $70 million.
* The diversion of 26 United Airlines flights to non-polar or less-than-optimum polar routes during several days of disturbed space weather in January 2005. The flights were diverted to avoid the risk of HF radio blackouts during PCA events. The increased flight time and extra landings and takeoffs required by such route changes increase fuel consumption and raise cost, while the delays disrupt connections to other flights.
* Disabling of the Federal Aviation Administration's recently implemented GPS-based Wide Area Augmentation System (WAAS) for 30 hours during the severe space weather events of October-November 2003.

With increasing awareness and understanding of space weather effects on their technologies, industries have responded to the threat of extreme space weather through improved operational procedures and technologies. As just noted, airlines re-route flights scheduled for polar routes during intense solar energetic particle events in order to preserve reliable communications. Alerted to an impending geomagnetic storm by NOAA's Space Weather Prediction Center (SWPC) and monitoring ground currents in real-time, power grid operators take defensive measures to protect the grid against geomagnetically induced currents (GICs). Similarly, under adverse space weather conditions, launch personnel may delay a launch, and satellite operators may postpone certain operations (e.g., thruster firings). For the spacecraft industry, however, the primary approach to mitigating the effects of space weather is to design satellites to operate under extreme environmental conditions to the maximum extent possible within cost and resource constraints. GPS modernization through the addition of two new navigation signals and new codes is expected to help mitigate space weather effects (e.g., ranging errors, fading caused by ionospheric scintillation), although to what degree is not known. These technologies will come on line incrementally over the next 15 years as new GPS satellites become operational. In the meantime, the Federal Aviation Administration will maintain "legacy" non-GPS-based navigation systems as a backup, while other GPS users (e.g., offshore drilling companies) can postpone operations for which precision position knowledge is required until the ionospheric disturbance is over.

The Collateral Impacts of Space Weather

Because of the interconnectedness of critical infrastructures in modern society, the impacts of severe space weather events can go beyond disruption of existing technical systems and lead to short-term as well as to long-term collateral socioeconomic disruptions. Electric power is modern society's cornerstone technology, the technology on which virtually all other infrastructures and services depend. Although the probability of a wide-area electric power blackout resulting from an extreme space weather event is low, the consequences of such an event could be very high, as its effects would cascade through other, dependent systems. Collateral effects of a longer-term outage would likely include, for example, disruption of the transportation, communication, banking, and finance systems, and government services; the breakdown of the distribution of potable water owing to pump failure; and the loss of perishable foods and medications because of lack of refrigeration. The resulting loss of services for a significant period of time in even one region of the country could affect the entire nation and have international impacts as well.

Extreme space weather events are low-frequency/high-consequence (LF/HC) events and as such present—in terms of their potential broader, collateral impacts—a unique set of problems for public (and private) institutions and governance, different from the problems raised by conventional, expected, and frequently experienced events.

As a consequence, dealing with the collateral impacts of LF/HC events requires different types of budgeting and management capabilities and consequently challenges the basis for conventional policies and risk management strategies, which assume a universe of constant or reliable conditions. Moreover, because systems can quickly become dependent on new technologies in ways that are unknown and unexpected to both developers and users, vulnerabilities in one part of the broader system have a tendency to spread to other parts of the system. Thus, it is difficult to understand, much less to predict, the consequences of future LF/HC events. Sustaining preparedness and planning for such events in future years is equally difficult.

Future Vulnerabilities

Our knowledge and understanding of the vulnerabilities of modern technological infrastructure to severe space weather and the measures developed to mitigate those vulnerabilities are based largely on experience and knowledge gained during the past 20 or 30 years, during such episodes of severe space weather as the geomagnetic superstorms of March 1989 and October-November 2003. As severe as some of these recent events have been, the historical record reveals that space weather of even greater severity has occurred in the past—e.g., the Carrington event of 1859[2] and the great geomagnetic storm of May 1921—and suggests that such extreme events, though rare, are likely to occur again some time in the future. While the socioeconomic impacts of a future Carrington event are difficult to predict, it is not unreasonable to assume that an event of such magnitude would lead to much deeper and more widespread socioeconomic disruptions than occurred in 1859, when modern electricity-based technology was still in its infancy.

A more quantitative estimate of the potential impact of an unusually large space weather event has been obtained by examining the effects of a storm of the magnitude of the May 1921 superstorm on today's electric power infrastructure. Despite the lessons learned since 1989 and their successful application during the October-November 2003 storms, the nation's electric power grids remain vulnerable to disruption and damage by severe space weather and have become even more so, in terms of both widespread blackouts and permanent equipment damage requiring long restoration times. According to a study by the Metatech Corporation, the occurrence today of an event like the 1921 storm would result in large-scale blackouts affecting more than 130 million people and would expose more than 350 transformers to the risk of permanent damage.

SPACE WEATHER INFRASTRUCTURE

Space weather services in the United States are provided primarily by NOAA's SWPC and the U.S. Air Force's (USAF's) Weather Agency (AFWA), which work closely together to address the needs of their civilian and military user communities, respectively. The SWPC draws on a variety of data sources, both space- and ground-based, to provide forecasts, watches, warnings, alerts, and summaries as well as operational space weather products to civilian and commercial users. Its primary sources of information about solar activity, upstream solar wind conditions, and the geospace environment are NASA's Advanced Composition Explorer (ACE), NOAA's GOES and POES satellites, magnetometers, and the USAF's solar observing networks. Secondary sources include SOHO and STEREO as well as a number of ground-based facilities. Despite a small and unstable budget (roughly $6 million to $7 million U.S. dollars annually) that limits capabilities, the SWPC has experienced a steady growth in customer base, even during the solar minimum years, when disturbance activity is lower. The focus of the USAF's space weather effort is on providing situational knowledge of the real-time space weather environment and assessments of the impacts of space weather on different Department of Defense missions. The Air Force uses NOAA data combined with data from its own assets

[2] The Carrington event is by several measures the most severe space weather event on record. It produced several days of spectacular auroral displays, even at unusually low latitudes, and significantly disrupted telegraph services around the world. It is named after the British astronomer Richard Carrington, who observed the intense white-light flare associated with the subsequent geomagnetic storm.

such as the Defense Meteorological Satellites Program satellites, the Communications/Navigation Outage Forecasting System, the Solar Electro-Optical Network, the Digital Ionospheric Sounding System, and the GPS network.

NASA is the third major element in the nation's space weather infrastructure. Although NASA's role is scientific rather than operational, NASA science missions such as ACE provide critical space weather information, and NASA's Living with a Star program targets research and technologies that are relevant to operations.

NASA-developed products that are candidates for eventual transfer from research to operations include sensor technology and physics-based space weather models that can be transitioned into operational tools for forecasting and situational awareness.

Other key elements of the nation's space weather infrastructure are the solar and space physics research community and the emerging commercial space weather businesses. Of particular importance are the efforts of these sectors in the area of model development.

Space Weather Forecasting: Capabilities and Limitations

One of the important functions of a nation's space weather infrastructure is to provide reliable long-term forecasts, although the importance of forecasts varies according to industry.[3] With long-term (1-to-3-day) forecasts and minimal false alarms,[4] the various user communities can take actions to mitigate the effects of impending solar disturbances and to minimize their economic impact. Currently, NOAA's SWPC can make probability forecasts of space weather events with varying degrees of success. For example, the SWPC can, with moderate confidence, predict the occurrence probability of a geomagnetic storm or an X-class flare 1 to 3 days in advance, whereas its capability to provide even short-term (less than 1 day) or long-term forecasts of ionospheric disturbances—information important for GPS users—is poor. The SWPC has identified a number of critical steps needed to improve its forecasting capability, enabling it, for example, to provide high-confidence long- and short-term forecasts of geomagnetic storms and ionospheric disturbances. These steps include securing an operational solar wind monitor at L1; transitioning research models (e.g., of coronal mass ejection propagation, the geospace radiation environment, and the coupled magnetosphere/ionosphere/atmosphere system) into operations, and developing precision GPS forecast and correction tools. The re-

[3] For the spacecraft industry, for example, space weather predictions are less important than knowledge of climatology and especially of the extremes within a climate record.

[4] False alarms are disruptive and expensive. Accurate forecasts of a severe magnetic storm would allow power companies to mitigate risk by canceling planned maintenance work, providing additional personnel to deal with adverse effects, and reducing the amount of power transfers between adjacent systems in the grid. However, as was pointed out during the workshop, if the warning proved to be a false alarm and planned maintenance was canceled, the cost of large cranes, huge equipment, and a great deal of material and manpower sitting idle would be very high.

quirement for a solar wind monitor at L1 is particularly important because ACE, the SWPC's sole source of real-time upstream solar wind and interplanetary magnetic field data, is well beyond its planned operational life, and provisions to replace it have not been made.

UNDERSTANDING THE SOCIETAL AND ECONOMIC IMPACTS OF SEVERE SPACE WEATHER

The title of the workshop on which this report is based, "The Societal and Economic Impacts of Severe Space Weather," perhaps promised more than this subsequent report can fully deliver. What emerged from the presentations and discussions at the workshop is that the invited experts understand well the effects of at least moderately severe space weather on specific technologies, and in many cases know what is required to mitigate them, whether enhanced forecasting and monitoring capabilities, new technologies (new GPS signals and codes, new-generation radiation-hardened electronics), or improved operational procedures. Limited information was also provided—and captured in this report—on the costs of space weather-induced outages (e.g., $50 million to $70 million to restore the $290 million Anik E2 to operational status) as well as of non-space-weather-related events that can serve as proxies for disruptions caused by severe space storms (e.g., $4 billion to $10 billion for the power blackout of August 2003), and an estimate of $1 trillion to $2 trillion during the first year alone was given for the societal and economic costs of a "severe geomagnetic storm scenario" with recovery times of 4 to 10 years.

Such cost information is interesting and useful—but as the outcome of the workshop and this report make clear, it is at best only a starting point for the challenge of answering the question implicit in the title: What are the societal and economic impacts of severe space weather? To answer this question quantitatively, multiple variables must be taken into account, including the magnitude, duration, and timing of the event; the nature, severity, and extent of the collateral effects cascading through a society characterized by strong dependencies and interdependencies; the robustness and resilience of the affected infrastructures; the risk management strategies and policies that the public and private sectors have in place; and the capability of the responsible federal, state, and local government agencies to respond to the effects of an extreme space weather event. While this workshop, along with its report, has gathered in one place much of what is currently known or suspected about societal and economic impacts, it has perhaps been most successful in illuminating the scope of the myriad issues involved, and the gaps in knowledge that remain to be explored in greater depth than can be accomplished in a workshop. A quantitative and comprehensive assessment of the societal and economic impacts of severe space weather will be a truly daunting task, and will involve questions that go well beyond the scope of the present report.

AMERICA'S STRATEGIC POSTURE

The Final Report of the Congressional Commission On the Strategic Posture of the United States (Excerpts)

Chaired by William J. Perry, Secretary of Defense

For President William J. Clinton

2009

Highlights:

"We note . . . that the United States has done little to reduce its vulnerability to attack with electromagnetic pulse weapons and recommend that current investments in modernizing the national power grid take account of this risk."[1]

Found online at http://media.usip.org/reports/strat_posture_report.pdf

EXCERPTS

ON PREVENTION AND PROTECTION: THE THREAT FROM ELECTROMAGNETIC PULSE WEAPONS[2]

The United States should take steps to reduce the vulnerability of the nation and the military to attacks with weapons designed to produce electromagnetic pulse (EMP) effects. We make this recommendation although the Commission is divided over how imminent a threat this is. Some commissioners believe it to be a high priority threat, given foreign activities and terrorist intentions. Others see it as a serious potential threat, given the high level of vulnerability. Those vulnerabilities are of many kinds. U.S. power projection forces might be subjected to an EMP attack by an enemy calculating — mistakenly — that such an attack would not involve risks of U.S. nuclear retaliation. The homeland might be attacked by terrorists or even state actors with an eye to crippling the U.S. economy and American society. From a technical perspective, it is possible that such attacks could have catastrophic consequences. For example, successful attacks could shut down the electrical system, disable the internet and computers and the economic activity on which they depend, incapacitate transportation systems (and thus the delivery of food and other goods), etc.

Prior commissions have investigated U.S. vulnerabilities and found little activity under way to address them. Some limited defensive measures have been ordered by the Department of Defense to give some protection to important operational communications. But EMP vulnerabilities have not yet been addressed effectively by the Department of Homeland Security. Doing so could take several years. The EMP commission has recommended numerous measures that would mitigate the damage that might be wrought by an EMP attack. The Stimulus Bill of February 9, 2009, allocates $11 billion to DOE for "for smart grid activities, including to modernize the electric grid." Unless such improvements in the electric grid are focused in part on reducing EMP vulnerabilities, vulnerability might well increase.

FINDINGS

The United States is highly vulnerable to attack with weapons designed to produce electromagnetic pulse effects.

RECOMMENDATIONS

EMP vulnerabilities should be reduced as the United States modernizes its electric power grid.

Metatech Corporation | Meta-R-323

Intentional Electromagnetic Interference (IEMI) and Its Impact on the U.S. Power Grid

William Radasky and Edward Savage

Produced for OAK RIDGE NATIONAL LABORATORY

2010

Highlights:

"It is clear that the biggest threat is against the civil infrastructure, shutting down the control electronics associated with the power grid, the telecom network or other parts of the critical infrastructure."

"The modern civil infrastructure is very dependent on computers, which operate at logic levels of a few volts. So an intentional interference can occur at a few volts in critical circuits, causing logic upset."

Found online at
http://www.ferc.gov/industries/electric/indus-act/reliability/cybersecurity/ferc_meta-r-323.pdf

INTRODUCTION

1.1 BACKGROUND

The term, "electromagnetic pulse" (EMP) has unfortunately been used in recent years (mainly by the media) to describe many different types of electromagnetic threats to electronic systems. In this report we will differentiate this general type of electromagnetic threat from the high-altitude electromagnetic pulse (HEMP). The HEMP is generated from a nuclear detonation in space, but the intense electromagnetic fields created there reach the Earth's surface. In the case of non-nuclear EMP, there are many subcategories of terms that describe this electromagnetic threat, which we will clarify and discuss in this report. In general we are speaking of the intense electromagnetic fields generated by a repeatable (non-explosive) high-power generator, which are directed to a target by an antenna. Our concern is how to protect our commercial infrastructure from these new mobile threats. We will refer specifically to this threat as IEMI (intentional electromagnetic interference).

In order to fully describe the terminology we will first describe the term "High Power Electromagnetics (HPEM)"; it has been used for many years and generally describes a set of transient EM environments where the peak electric and magnetic fields can be very high. The typical environments considered are the electromagnetic fields from nearby lightning strikes, the electromagnetic fields near an electrostatic discharge, the electromagnetic fields created in substations due to switching and arcing events, and the electromagnetic fields created by radar systems. In addition to these natural and accidental EM threats, we add, the electromagnetic pulse (HEMP) created by high altitude nuclear bursts and the intentional electromagnetic interference (IEMI).

Figure 1-1 shows qualitatively several of these electromagnetic environments, along with the narrowband and wideband IEMI threats that are the subject of this report. It should be noted that the EMC Society of the IEEE has a technical committee TC-5 with the title of "High Power Electromagnetics" dealing with all of these subjects. In addition, the IEC is developing standards to protect commercial equipment and systems under Subcommittee 77C, which is entitled "EMC: High power transient phenomena".

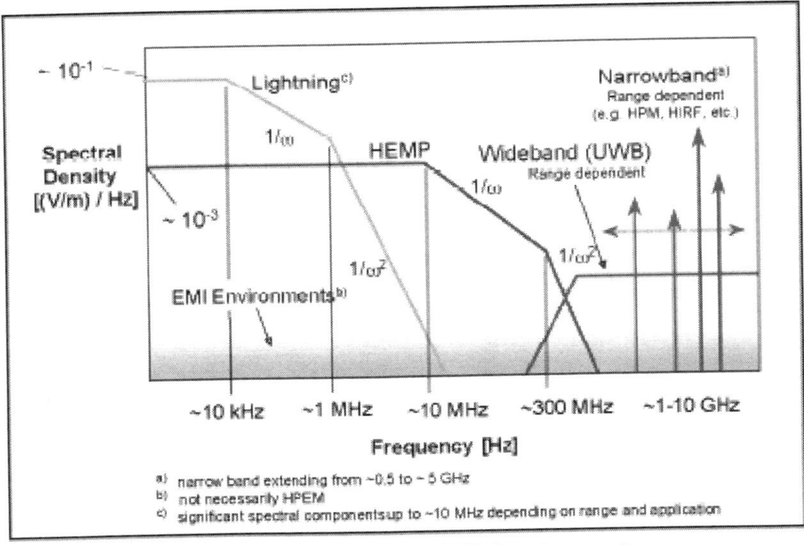

Figure 1-1. Comparison of HPEM environments.

Most recently two new terms have arisen in the EMC field – EM Terrorism[3] and Intentional Electromagnetic Interference (IEMI).[4] Over the past 10 years the scientific community has decided to accept the more generic term IEMI, which includes EM Terrorism. In February 1999 at a workshop held at the Zurich EMC Symposium, a widely accepted definition for IEMI was suggested: "Intentional malicious generation of electromagnetic energy introducing noise or signals into electric and electronic systems, thus disrupting, confusing or damaging these systems for terrorist or criminal purposes".

Note that hackers are not mentioned explicitly in this definition, although in most countries of the world, an attack on commercial interests for "entertainment" is also against the law. While the motives of the attackers may vary, the results can be the same for civil society. The scientific community has been working to understand this threat and to protect against it in a more precise manner.

While this report aims to inform the reader about the threat of IEMI against commercial electronic equipment and systems in general, it is clear that the biggest threat is against the civil infrastructure, as shutting down the control electronics associated with the power grid, the telecom network or other parts of the critical infrastructure could have widespread impacts.

1.2 PAST EXPERIENCE WITH HPEM EFFECTS ON SYSTEMS

While concern is often directed at modern electronic devices with solid-state digital electronics that are common today, damage to electronic systems has occurred in the past. In particular, in 1967, the USS Forrestal was involved in one of the worst cases of EMI ever documented. While sitting on the deck, a military aircraft was exposed to the ship's radar and accidentally fired its munitions, hitting another fully armed and fueled aircraft on the deck. The explosions and resulting fire caused severe damage to the carrier and resulted in 134 deaths. A later investigation discovered that a degraded cable shield termination on the first aircraft was the cause of the accident.[5]

Such occurrences of accidental EMI are not limited to the military. When anti-lock braking systems (ABS) were first introduced, problems arose in Germany on the autobahn when brakes were applied when the autos passed a nearby radio transmitter. This problem was mitigated by the placement of mesh screen.[6]

The medical care industry has also been affected by EMI. A 93-year-old heart attack victim died when the attached monitor and defibrillator shut down every time the radio transmitter was used in an ambulance. This was due to the metal fiberglass ambulance roof that allowed high levels of radiated radio fields inside the patient area of the ambulance.[7]

These instances of high-power electromagnetic (HPEM) fields impacting electrical systems were inadvertent consequences of a poor system design or implementation, abnormally large EM fields, or both. It is possible, however, to envision the use of HPEM sources to intentionally cause upset or damage in a system. Such a situation could occur in a military setting, where the HPEM environment could be directed towards an enemy system. More to the point for our concerns for civil society, an attack by hackers, criminals, or terrorists could produce IEMI.

IEMI concerns have been the subject of technical sessions in recent scientific symposia[8][9][10][11] and continue to be discussed in the popular press.[12][13] Although there are several unconfirmed accounts of instances where such (EM) weapons have been used against civil and military systems[14][15], obtaining clear, convincing and documented evidence of these cases remains elusive.

While there is a lack of clear proof linking the use of such HPEM sources to attack civil facilities, several governments have publicly indicated that they are assessing the possible effects of HPEM environments on their systems and infrastructure. Two examples include a research effort in Sweden[16] and recent testimony before the U.S. Congress about the possibility of the use of radio frequency (RF) weapons.[17]

1.3 IMPACTS OF IEMI ON SOCIETY

The first question one might ask is whether there really is any reason for society to be concerned about this problem. In fact there are many as indicated below:

- Terrorist threats are increasing world-wide
- Covert operation outside physical barriers are attractive
- Technological advances have produced higher-energy RF sources and more efficient antennas
- Proliferation of IEMI sources is increasing
- Society's dependence on information and on automated mission-critical

and safety-critical electronic systems is increasing

* EM susceptibility of new high density IT systems working at higher frequencies and lower voltages is increasing

In August 1999 this problem was recognized by the International Radio Scientific Union (URSI) during a special session that resulted in an URSI resolution. The URSI "Resolution of Criminal Activities using Electromagnetic Tools"[18] was intended to make people aware of:

* The existence of criminal activities using electromagnetic tools and associated phenomena
* The fact that criminal activities using electromagnetic tools can be undertaken covertly and anonymously and that physical boundaries such as fences and walls can be penetrated by electromagnetic fields
* The potentially serious nature of the effects of criminal activities using electromagnetic tools on the infrastructure and important functions in society such as transportation, communication, security, and medicine
* That the possible disruptions of the health and economic activities of nations could have major consequences
* The URSI Council recommended to the scientific community in general, and the EMC community in particular, to take account of this threat and to undertake the following actions:
    - Perform additional research pertaining to criminal activities using electromagnetic tools in order to establish appropriate levels of vulnerability
    - Investigate techniques for appropriate protection against criminal activities using electromagnetic tools and to provide methods that can be used to protect the public from the damage that can be done to the infrastructure by terrorists
    - Develop high-quality testing and assessment methods to evaluate system performance in these special electromagnetic environments
    - Provide data regarding the formulation of standards of protection and support standardization work

It is noted that the International Electrotechnical Commission (IEC) added the IEMI threat to its previous standardization work dealing with HEMP in 1999.

High-Impact, Low-Frequency Event Risk to the North American Bulk Power System

A Jointly-Commissioned Summary Report of the North American Electric Reliability Corporation and the U.S. Department of Energy's November 2009 Workshop

2010

Highlights:

"A class of risks, called High-Impact, Low-Frequency (HILF) events, has recently become a renewed focus of risk managers and policy makers. These risks have the potential to cause catastrophic impacts on the electric power system, but either rarely occur, or, in some cases, have never occurred."

"Examples of HILF risks include coordinated cyber, physical, and blended attacks, the high-altitude detonation of a nuclear weapon, and major natural disasters like earthquakes, tsunamis, large hurricanes, pandemics, and geomagnetic disturbances caused by solar weather."

Found online at http://www.ourenergypolicy.org/wp-content/uploads/2012/05/HILF.pdf

EXECUTIVE SUMMARY

The bulk power system is one of North America's most critical infrastructures, underpinning the continent's governments, economy and society. As reliance on electricity-dependent technology has increased, the reliability of the power grid has become more important each day. The electric sector has recognized the importance of the infrastructure it operates and has had a long history of successfully managing day-to-day operational and probabilistic risk to the reliability of the system to ensure the "lights stay on" for consumers.

A class of risks, called High-Impact, Low-Frequency (HILF) events, has recently become a renewed focus of risk managers and policy makers. These risks have the potential to cause catastrophic impacts on the electric power system, but either rarely occur, or, in some cases, have never occurred. Examples of HILF risks include coordinated cyber, physical, and blended attacks, the high-altitude detonation of a nuclear weapon, and major natural disasters like earthquakes, tsunamis, large hurricanes, pandemics, and geomagnetic disturbances caused by solar weather. HILF events truly transcend other risks to the sector due to their magnitude of impact and the relatively limited operational experience in addressing them. Deliberate attacks (including acts of war, terrorism, and coordinated criminal activity) pose especially unique scenarios due to their inherent unpredictability and significant national security implications. As concerns over these risks have increased, the electric sector is working to take a leadership position among other Critical Infrastructure and Key Resource (CIKR) sectors in addressing these risks.

THE HIGH-IMPACT, LOW-FREQUENCY (HILF) EVENT RISK EFFORT

To facilitate the development of a sector-wide roadmap for further public/private collaboration on these issues, the North American Electric Reliability Corporation (NERC) and U.S. Department of Energy (DOE) jointly sponsored a workshop on HILF risks in November, 2009. The approximately 110 attendees at the closed session included representatives from the U.S.'s Congressional Staff, Department of Defense (DOD), Department of Homeland Security (DHS), DOE, Department of Health and Human Services (HHS), EMP Commission, and Federal Energy Regulatory Commission (FERC). Representatives from each of the North American electric industry's major sectors, including investor owned utilities, cooperatives, and municipal utilities were also in attendance, as were many risk experts.

This report is intended to summarize the proceedings and discussions at the two-day session. Proposals for action and mitigating options discussed herein reflect the thoughts of the session participants, and, while they may represent a largely consensus-based view, they are not intended to be conclusive or exhaustive. Most of the proposals in this document identify areas where further work is needed and provide initial guidance on the kinds of efforts that must be undertaken.

As these proposals for action are considered, it is important to place HILF risks in context of the larger landscape of risk and concerns facing the electric sector over the coming years. NERC's 2009 Long-Term Reliability Assessment[19], for example, identified nine emerging issues expected to impact reliability by 2018 including climate legislation, smart grid, cyber security, transmission siting, variable generation issues, workforce issues, and reactive power. Several of these are reflective of other legislative and regulatory priorities. In addition, the sector is expected to require significant infrastructure additions[20] to meet demand as economic recovery continues over the coming years.

ADDRESSING HILF RISK

The interconnected and interdependent nature of the bulk power system requires that risk management actions be consistently and systematically applied across the entire system to be effective. The magnitude of such an effort should not be underestimated. The North American bulk power system is comprised of more than 200,000 miles of high-voltage transmission lines, thousands of generation plants, and millions of digital controls.[21] More than 1,800 entities own and operate portions of the system, with thousands more involved in the operation of distribution networks across North America. These entities range in size from large investor-owned utilities with over 20,000 employees to small cooperatives with only ten. The systems and facilities comprising the larger system have differing configurations, design schemes, and operational concerns. Referring to any mitigation on such a system as "easily-deployed," "inexpensive," or "simple" is an inaccurate characterization of the work required to implement these changes.

As mitigating options are further considered, it is also important to note that it is impossible to fully protect the system from every threat or threat actor. Sound management of these and all risks to the sector must take a holistic approach, with specific focus on determining the appropriate balance of resilience, restoration, and protection. A successful risk management approach will begin by identifying the threat environment and protection goals for the system, balancing expected outcomes against the costs associated with proposed mitigations.

This balance must be carefully considered with input from both electric sector and government authorities. Building on the inherent resilience of the system and enhancing the response of the system as a whole to unconventional stresses should be a cornerstone of these efforts. Determining appropriate cost ceilings and recovery mechanisms for protections related to HILF risks will be critical to ensuring a viable approach to addressing them. The electricity industry and government authorities must also coordinate to improve two-way information sharing and communication practices relative to HILF risks. The sector is heavily reliant on information from the public sector for each risk discussed in this document.

Common elements of addressing HILF risk must also include a focus on raising awareness across the sector and creating opportunities to discuss specific issues in technical detail. In many cases, this will take the form of creating various task forces designed to bring together personnel from the risk community, electric sector, government, and equipment manufacturers. These task forces will provide a comprehensive view of technical implications and potential solutions to the challenges posed by these risks.

Additional research and development will also be needed in certain areas to ensure mitigating technology solutions are available to industry. This is particularly important with reference to cyber security and electro-magnetic pulse threats. Ensuring protections can be built-in to future products as opposed to being delivered as a "bolt-on" retrofit will greatly improve the cost-effectiveness of protections on a going-forward basis. Hardening of existing assets will also be important, as many assets have long life cycles.

HILF RISK DISCUSSED IN THIS REPORT

While HILF risks can include other extreme events like major natural disasters, meteor strikes, and deliberate attacks or acts of war, the November workshop focused on three specific threats as identified by the HILF Steering Committee in the planning process: Coordinated Cyber/Physical Attack, Pandemic Illness, and Geomagnetic and Electromagnetic Events. Each section identifies the threat to the system, the system's vulnerabilities, and the consequences that could occur were these vulnerabilities to be exploited. This discussion is followed by a consideration of various mitigating options and *Proposals for Action.*

Highlights: Coordinated Attack Risk

The risk of a coordinated cyber, physical, or blended attack against the North American bulk power system has become more acute over the past 15 years as digital communicating equipment has introduced cyber vulnerability to the system, and resource optimization trends have allowed some inherent physical redundancy within the system to be reduced. The specific concern with respect to these threats is the targeting of multiple key nodes on the system that, if damaged, destroyed, or interrupted in a coordinated fashion, could bring the system outside the protection provided by traditional planning and operating criteria. Such an attack would behave very differently than traditional risks to the system in that an intelligent attacker could mount an adaptive attack that would manipulate assets and potentially provide misleading information to system operators attempting to address the issue. While no such attack has occurred on the bulk power system to date, the electric sector has taken important steps toward mitigating these issues with the development of NERC's Critical Infrastructure Protection standards[22], the standing Critical Infrastructure Protection Committee[23], and a myriad of other efforts. More comprehensive work is needed, however, to realize the vision of a secure grid. Better technology solutions for the cyber portion of the threat should be developed, with specific focus on forensic tools and network architectures to support graceful system degradation that would allow operators to "fly with fewer controls." Component and system design criteria should also be reevaluated with respect to these threats and an eye toward designing for survivability. Prioritization of key assets for protection will be a critical component of a successful mitigation approach.

Highlights: Pandemic Risk

Pandemic risk differs from many of the other threats facing the system in that it is a "people event." The principal vulnerability with respect to a pandemic is the loss of staff critical to operating the electric power system. Without these personnel, operational issues on the system would increase as less-trained or less-experienced individuals work to operate generation plants, address mechanical failures, restore power following outages caused by weather and other natural events, and operate the system. The sector recently experienced a mild pandemic through the 2009 A/H1N1 outbreak. This pandemic's effects on society were very limited and are not representative of the scenarios of concern to the electric sector. While many entities within the sector have developed advanced pandemic plans, the sector is ultimately reliant on government health authorities for quality and timely information on the spread and severity of a pandemic. Clear triggers from these authorities are needed for the sector to make appropriate response decisions in the event of a severe outbreak.

Highlights: Geomagnetic Disturbances, High Altitude Electromagnetic Pulse Events, and Intentional Electromagnetic Interference Threats

Geomagnetic disturbances, the earthly effects of solar weather, are not a new threat to the electric sector. Recent analysis by Metatech and Storm Analysis Consultants[24,25,26,27] suggests, however, that the potential extremes of the geomagnetic threat environment may be much greater than previously anticipated. Geomagnetically-induced currents on system infrastructure have the potential to result in widespread tripping of key transmission lines and irreversible physical damage to large transformers.[28,29,30,31] The 1989 event that caused a blackout of the Hydro Québec system provided important lessons to the sector. Since that time, the sector has adopted operational procedures to reduce the vulnerability to geomagnetic storms and has installed certain protections in areas most prone to impact as recommended by Oak Ridge National Labs in their report on the March 1989 event.[32] More work is needed, however, to consider the potential impacts larger storms may have and develop viable, cost-effective mitigations, potentially at lower geographic latitudes than previously thought necessary.

The high-altitude detonation of a large nuclear device or other electromagnetic weapon could have devastating effects on the electric sector, interrupting system operation and potentially damaging many devices simultaneously. A coordinated attack involving intentional electromagnetic interference (IEMI) could result in more localized and targeted impacts that may also cause significant impacts to the sector.

The physical damage of certain system components (e.g. extra-high-voltage transformers) on a large scale, as could be effected by any of these threats, could result in prolonged outages as procurement cycles for these components range from months to years. Many of these components are manufactured overseas, with little manufacturing capability remaining in North America. The impacts of these events on the power system are not yet fully understood across the sector and warrant further collaborative work to identify the prioritized "top ten" mitigation steps that are both cost-effective and sufficient to protect the power system from the widespread catastrophic damage that could result from any of these events.

NEXT STEPS

The *Proposals for Action* outlined in this report are intended to provide input into a formal action plan to address these issues. They do not, in and of themselves, constitute this plan. The effort needed to address these risks will require intense coordination and a significant resource commitment from all entities involved. The time needed to address these issues and complete the work contemplated herein will be measured in years. NERC and the U.S. DOE will work together with the electric sector, manufacturers, and other government authorities to support the development and execution of a clear and concise action plan to ensure accountability and coordinated action on these issues going forward.

Large Power Transformers and the U.S. Electric Grid

Infrastructure Security and Energy Restoration Office of Electricity Delivery and Energy Reliability U.S. Department of Energy

United States of America Department of Energy

2012

Highlights:

"Large Power Transformers (LPTs) are custom-designed equipment that entail significant capital expenditures and long lead times due to an intricate procurement and manufacturing process."

"Because LPTs are very expensive and tailored to customers' specifications, they are usually neither interchangeable with each other nor produced for extensive spare inventories."

"The average lead time for manufacture of an LPT is between five and 16 months; however, the lead time can extend beyond 20 months if there are any supply disruptions or delays with the supplies, raw materials, or key parts."

"The United States has limited production capability to manufacture LPTs."

Found online at
http://energy.gov/sites/prod/files/Large%20Power%20Transformer%20Study%20-%20June%202012_0.pdf

EXECUTIVE SUMMARY

The Office of Electricity Delivery and Energy Reliability, U.S. Department of Energy (DOE) assessed the procurement and supply environment of large power transformers (LPT)[5] in this report. LPTs have long been a major concern for the U.S. electric power sector, because failure of a single unit can cause temporary service interruption and lead to collateral damages, and it could be difficult to quickly replace it. Key industry sources—including the *Energy Sector Specific Plan*, the National Infrastructure Advisory Council's *A Framework for Establishing Critical Infrastructure Resilience Goals* and the North American Electric Reliability Corporation's *Critical Infrastructure Strategic Roadmap*—have identified the limited availability of spare LPTs as a potential issue for critical infrastructure resilience in the United States, and both the public and private sectors have been undertaking a variety of efforts to address this concern. Therefore, DOE examined the following topics in this report: characteristics and procurement of LPTs, including key raw materials and transportation; historical trends and future demands; global and domestic LPT suppliers; and potential issues in the global sourcing of LPTs.

LPTs are custom-designed equipment that entail significant capital expenditures and long lead times due to an intricate procurement and manufacturing process. Although the costs and pricing vary by manufacturer and by size, an LPT can cost millions of dollars and weigh between approximately 100 and 400 tons (or between 200,000 and 800,000 pounds). Procurement and manufacturing of LPTs is a complex process that requires prequalification of manufacturers, a competitive bidding process, the purchase of raw materials, and special modes of transportation due to its size and weight. The result is the possibility of extended lead times that could stretch beyond 20 months if the manufacturer has difficulty obtaining certain key parts or materials. Two raw materials—copper and electrical steel—account for over 50 percent of the total cost of an LPT. Electrical steel is used for the core of a power transformer and is critical to the efficiency and performance of the equipment; copper is used for the windings. In recent years, the price volatility of these two commodities in the global market has affected the manufacturing conditions and procurement strategy for LPTs.

The rising global demand for copper and electrical steel can be partially attributed to the increased power and transmission infrastructure investment in growing economies as well as the replacement market for aging infrastructure in developed countries. The United States is one of the world's largest markets for power transformers and holds the largest installed base of LPTs—and this installed base is aging. The average age of installed LPTs in the United States is approximately 40 years, with 70 percent of LPTs being 25 years or older. While the life expectancy of a power transformer varies depending on how it is used, aging power transformers

[5] Throughout this report, the term large power transformer (LPT) is broadly used to describe a power transformer with a maximum capacity rating greater and or equal to 100 MVA unless otherwise noted.

are subject to an increased risk of failure.

Since the late 1990's, the United States has experienced an increased demand for LPTs; however, despite the growing need, the United States has had a limited domestic capacity to produce LPTs. In 2010, six power transformer-manufacturing facilities existed in the United States, and together, they met approximately 15 percent of the Nation's demand for power transformers of a capacity rating greater than or equal to 60 MVA. Although the exact statistics are unavailable, global power transformer supply conditions indicate that the Nation's reliance on foreign manufacturers is even greater for extra high-voltage (EHV) power transformers with a maximum voltage rating greater than or equal to 345 kV.

However, the domestic production capacity for LPTs in the United States is improving. In addition to EFACEC's first U.S. transformer plant that began operation in Rincon, Georgia in April 2010, at least three new or expanded facilities will produce EHV LPTs starting in 2012 and beyond. These include: SPX Transformer Solution's facility in Waukesha, Wisconsin, which completed expansion in April 2012; Hyundai Heavy Industries' new manufacturing facility, which was inaugurated in Montgomery, Alabama in November 2011; and Mitsubishi's proposed development of a power transformer plant in Memphis, Tennessee, which is expected to be completed in 2013.

The upward trend of transmission infrastructure investment in the United States since the late 1990s is one of the key drivers for the recent addition of domestic manufacturing capacity for power transformers. Between 2005 and 2011, the total value of LPTs imported to the United States grew by 188 percent (or at an annual growth rate of 34 percent) from $284 to $817 million U.S. dollar. Power transformers are globally traded equipment, and the demand for this machinery is forecasted to continue to grow at a compound annual growth rate of three to seven percent in the United States according to industry sources. In addition to replacing aging infrastructure, the United States has needs for transmission expansion and upgrades to accommodate new generation connections and maintain electric reliability.

While global procurement has become a common practice for many utilities to meet their growing need for LPTs, there are several challenges associated with it. Such challenges include: the potential for extended lead times due to unexpected global events (e.g., hurricanes) or difficulty in transportation; the fluctuation of currency exchange rates and material prices; and cultural differences and communication barriers. The utility industry is also facing the challenge of maintaining experienced in-house workforce that is able to address procurement and maintenance issues.

The U.S. electric power grid is one of the Nation's critical life-line infrastructure on which many other critical infrastructure depend, and the destruction of this infrastructure can cause a significant impact to national security and the U.S. economy. The U.S. electric grid faces a wide variety of possible threats, including natural, physical, cyber, and space weather. While the potential effect of these threats on the electric power grid is uncertain, public and private stakeholders of the energy industry are considering various risk management strategies to mitigate potential impacts. This DOE report, through the assessment of LPT procurement and supply issues, provides information to help the industry's continuous efforts to build critical energy infrastructure resilience in today's complex, interdependent global economy.

United States Government Accountability Office

Cybersecurity: Challenges in Securing the Electricity Grid

Testimony Before the Committee on Energy
and Natural Resources, U.S. Senate

Statement of Gregory C. Wilshusen
Director Information Security Issues

2012

Highlights:

"In testimony, the Director of National Intelligence noted a dramatic increase in cyber activity targeting U.S. computers and systems, including a more than tripling of the volume of malicious software....The electricity grid's reliance on IT systems and networks exposes it to potential and known cybersecurity vulnerabilities, which could be exploited by attackers."

"The potential impact of such attacks has been illustrated by a number of recently reported incidents and can include fraudulent activities, damage to electricity control systems, power outages, and failures in safety equipment."

Found online at http://www.gao.gov/assets/600/592508.pdf

WHAT GAO FOUND

The threats to systems supporting critical infrastructures are evolving and growing. In testimony, the Director of National Intelligence noted a dramatic increase in cyber activity targeting U.S. computers and systems, including a more than tripling of the volume of malicious software. Varying types of threats from numerous sources can adversely affect computers, software, networks, organizations, entire industries, and the Internet itself. These include both unintentional and intentional threats, and may come in the form of targeted or untargeted attacks from criminal groups, hackers, disgruntled employees, nations, or terrorists. The interconnectivity between information systems, the Internet, and other infrastructures can amplify the impact of these threats, potentially affecting the operations of critical infrastructures, the security of sensitive information, and the flow of commerce. Moreover, the electricity grid's reliance on IT systems and networks exposes it to potential and known cybersecurity vulnerabilities, which could be exploited by attackers. The potential impact of such attacks has been illustrated by a number of recently reported incidents and can include fraudulent activities, damage to electricity control systems, power outages, and failures in safety equipment.

To address such concerns, multiple entities have taken steps to help secure the electricity grid, including the North American Electric Reliability Corporation, the National Institute of Standards and Technology (NIST), the Federal Energy Regulatory Commission, and the Departments of Homeland Security and Energy. These include, in particular, establishing mandatory and voluntary cybersecurity standards and guidance for use by entities in the electricity industry. For example, the North American Electric Reliability Corporation and the Federal Energy Regulatory Commission, which have responsibility for regulation and oversight of part of the industry, have developed and approved mandatory cybersecurity standards and additional guidance. In addition, NIST has identified cybersecurity standards that support smart grid interoperability and has issued a cybersecurity guideline. The Departments of Homeland Security and Energy have also played roles in disseminating guidance on security practices and providing other assistance.

As GAO previously reported, there were a number of ongoing challenges to securing electricity systems and networks. These include:

* A lack of a coordinated approach to monitor industry compliance with voluntary standards.
* Aspects of the current regulatory environment made it difficult to ensure the cybersecurity of smart grid systems.
* A focus by utilities on regulatory compliance instead of comprehensive security.
* A lack of security features consistently built into smart grid systems.
* The electricity industry did not have an effective mechanism for sharing information on cybersecurity and other issues.
* The electricity industry did not have metrics for evaluating cybersecurity.

Chairman Bingaman, Ranking Member Murkowski, and Members of the Committee:

Thank you for the opportunity to testify at today's hearing on the status of actions to protect the electricity grid from cyber attacks.

As you know, the electric power industry is increasingly incorporating information technology (IT) systems and networks into its existing infrastructure (e.g., electricity networks including power lines and customer meters). This use of IT can provide many benefits, such as greater efficiency and lower costs to consumers. Along with these anticipated benefits, however, cybersecurity and industry experts have expressed concern that, if not implemented securely, modernized electricity grid systems will be vulnerable to attacks that could result in widespread loss of electrical services essential to maintaining our national economy and security.

In addition, since 2003 we have identified protecting systems supporting our nation's critical infrastructure (which includes the electricity grid) as a government-wide high-risk area, and we continue to do so in the most recent update to our high-risk list.[6]

In my testimony today, I will describe cyber threats facing cyber-reliant critical infrastructures,[7] which include the electricity grid, and actions taken and challenges remaining to secure the grid against cyber attacks. In preparing this statement in July 2012, we relied on our previous work in this area, including studies examining efforts to secure the electricity grid and associated challenges and cybersecurity guidance.[8] (Please see the related GAO products in appendix I.) The products upon which this statement is based contain detailed overviews of the scope of our reviews and the methodology we used. We also reviewed documents from the Federal Energy Regulatory Commission, the North American Electric Reliability Corporation, the Department of Energy, including its Office of the Inspector General, and the Department of Homeland Security Industrial Control Systems Cyber Emergency Response Team, as well as publicly available reports on cyber incidents. The work on which this statement is based was performed in accordance with generally accepted government auditing standards. Those standards require that we plan and

[6] GAO's biennial high-risk list identifies government programs that have greater vulnerability to fraud, waste, abuse, and mismanagement or need transformation to address economy, efficiency, or effectiveness challenges. We have designated federal information security as a government wide high-risk area since 1997; in 2003, we expanded this high-risk area to include protecting systems supporting our nation's critical infrastructure—referred to as cyber-critical infrastructure protection, or cyber CIP. See, most recently, GAO, High-Risk Series: An Update, (Please see the related GAO products in appendix I.) The products upon which this GAO-11-278 (Washington, D.C.: February 2011).

[7] Federal policy established 18 critical infrastructure sectors. These include, for example, banking and finance, communications, public health, and energy. The energy sector includes subsectors for oil and gas and for electricity.

[8] GAO, Critical Infrastructure Protection: Cybersecurity Guidance Is Available, but More Can Be Done to Promote Its Use, GAO-12-92 (Washington, D.C.: Dec. 9, 2011), and Electricity Grid Modernization: Progress Being Made on Cybersecurity Guidelines, but Key Challenges Remain to be Addressed, GAO-11-117 (Washington, D.C.: Jan. 12, 2011).

perform audits to obtain sufficient, appropriate evidence to provide a reasonable basis for our findings and conclusions. We believe that the evidence obtained provided a reasonable basis for our findings and conclusions based on our audit objectives.

Terrorism and the Electric Power Delivery System

National Research Council of the National Academies

Committee on Enhancing the Robustness and Resilience of Future Electrical Transmission and Distribution in the United States to Terrorist Attack

Board on Energy and Environmental Systems
Division on Engineering and Physical Sciences

2012

Highlights:

"The electric power delivery system that carries electricity from large central generators to customers could be severely damaged by a small number of well-informed attackers. The system is inherently vulnerable because transmission lines may span hundreds of miles, and many key facilities are unguarded."

"Terrorist attacks on multiple-line transmission corridors could cause cascading blackouts. High-voltage transformers are of particular concern because they are vulnerable to attack, both from within and from outside the substation where they are located. These transformers are very large, difficult to move, custom-built, and difficult to replace. Most are no longer made in the United States, and the delivery time for new ones can run to months or years."

Found online at http://www.nap.edu/catalog.php?record_id=12050

SUMMARY

The electric power delivery system that carries electricity from large central generators to customers could be severely damaged by a small number of well-informed attackers. The system is inherently vulnerable because transmission lines may span hundreds of miles, and many key facilities are unguarded. This vulnerability is exacerbated by the fact that the power grid, most of which was originally designed to meet the needs of individual vertically integrated utilities, is now being used to move power between regions to support the needs of new competitive markets for power generation. Primarily because of ambiguities introduced as a result of recent restructuring of the industry and cost pressures from consumers and regulators, investment to strengthen and upgrade the grid has lagged, with the result that many parts of the bulk high-voltage system are heavily stressed.

A terrorist attack on the power system would lack the dramatic impact of the attacks in New York, Madrid, or London. It would not immediately kill many people or make for spectacular television footage of bloody destruction. But if it were carried out in a carefully planned way, by people who knew what they were doing, it could deny large regions of the country access to bulk system power for weeks or even months. An event of this magnitude and duration could lead to turmoil, widespread public fear, and an image of helplessness that would play directly into the hands of the terrorists. If such large extended outages were to occur during times of extreme weather, *they could also result in hundreds or even thousands of deaths due to heat stress or extended exposure to extreme cold.*

The largest power system disruptions experienced to date in the United States have caused high economic impacts. Considering that a systematically designed and executed terrorist attack could cause disruptions that were even more widespread and of longer duration, it is no stretch of the imagination to think that such attacks could entail costs of hundreds of billions of dollars-that is, perhaps as much as a few percent of the U.S. gross domestic product (GDP), which is currently about $12.5 trillion.

Electric systems are not designed to withstand or quickly recover from damage inflicted simultaneously on multiple components. Such an attack could be carried out by knowledgeable attackers with little risk of detection or interdiction. Further well-planned and coordinated attacks by terrorists could leave the electric power system in a large region of the country at least partially disabled for a very long time. Although there are many examples of terrorist and military attacks on power systems elsewhere in the world, to date international terrorists have shown limited interest in attacking the U.S. power grid. However, that should not be a basis for complacency. Since all parts of the economy, as well as human health and welfare, depend on electricity, the results could be devastating.

This report focuses on measures that could:

1. Make the power delivery system less vulnerable to attacks,
2. Restore power faster after an attack,

3. Make critical services less vulnerable while the delivery of conventional electric power has been disrupted.

THE NATURE OF THE PROBLEM

The U.S. power delivery system is remarkably complex. It is a network of substations, transmission lines, distribution lines, and other components that people can see as they drive around the country; it also includes the less visible devices that sense and report on the state of the system, the automatic and human controls that operate the system, and the intricate web of computers and communication systems that tie everything together. Enormous complexity and diversity also characterize the organizations and human systems that operate and manage the power delivery system. That complexity and diversity have become even greater in recent years as some parts of the system have been restructured while others have not, and as the role of state and federal regulators and other oversight bodies has shifted.

Today most power is generated by large central generating stations that are located far from the customers they serve. Transformers increase the voltage so that it can be carried efficiently over long distances. Substations then reduce the voltage and carry the power into the distribution network for delivery to customers.[33] Unlike trains or natural gas in pipelines, electric power cannot simply be sent via specific lines wherever dispatchers choose. Current flows through the system according to a set of physical laws. The system must be continually adjusted to keep all parts synchronized and in electrical balance. If corrections are not made immediately when imbalances occur, the result can be oscillations and other disturbances in the system that can result in a cascading failure over a wide area, as happened in the Northeast blackout of 2003.

Recent years have witnessed dramatic organizational changes in the U.S. electric power system. In some states, traditional vertically integrated companies that owned and operated the entire system from the generators to the customers' meters have been restructured in an effort to introduce competition. However, a few states are trying to undo some of the changes' and some states may never restructure. The push by federal regulators to introduce competition in bulk power across the country also has resulted in the transmission network being used in ways for which it was not designed. There have also been shifts in the relative responsibility of state and federal regulators.

Largely as a consequence of the uncertainties introduced by these changes, incentives for investment by private firms have become mixed, with the result that the physical capabilities of much of the transmission network have not kept pace with the increasing burden that is being placed on it. Other trends are more promising. The Energy Policy Act of 2005 includes provisions to strengthen the electric grid, including provisions for the introduction of mandatory reliability standards. Although not aimed specifically at protecting the grid against terrorism, the activities initiated under this statute will-if implemented-lead to a more robust transmission system that will be better able to withstand major disruptions.

Physical Vulnerability

Disruption in the supply of electric power can result from problems in any part of the system. The primary concern of this report is with power delivery. Substations and the large high-voltage transformers they contain are especially vulnerable, as are some transmission lines where the destruction of a small number of towers could bring down many kilometers of line. Terrorist attacks on multiple-line transmission corridors could cause cascading blackouts.

High-voltage transformers are of particular concern because they are vulnerable to attack, both from within and from outside the substation where they are located. These transformers are very large, difficult to move, custom-built, and difficult to replace. Most are no longer made in the United States, and the delivery time for new ones can run to months or years. The industry has made some progress toward building an inventory of spares, but these efforts could be overwhelmed by a large attack. Although easier to move and replace, other large components, such as high-voltage circuit breakers, are also a concern.

These problems are exacerbated by the current state of the transmission grid. It is aging and increasingly stressed, leaving it especially vulnerable to multiple failures following an attack. Many important pieces of equipment are decades old and lack improved technology that could help limit outages.

Cyber Vulnerability

Modern power systems rely heavily on automation, centralized control of equipment, and high-speed communications. The most critical systems are the supervisory control and data acquisition (SCADA) systems that gather real-time measurements from substations and send out control signals to equipment, such as circuit breakers. The many other control systems, such as substation automation or protection systems, can each only control local equipment. Other online computer systems, such as energy management systems (which analyze the reliability of the system against contingencies) or market systems (which manage the buying and selling of electricity), have only an indirect impact on the grid. But all such systems are potentially vulnerable to cyber attacks, whether through Internet connections or by direct penetration at remote sites. Any telecommunication link that is even partially outside the control of the system operators is a potentially insecure pathway into operations and a threat to the grid.

If they could gain access, hackers could manipulate SCADA systems to disrupt the flow of electricity, transmit erroneous signals to operators, block the flow of vital information, or disable protective systems. Cyber attacks are unlikely to cause extended outages, but if well coordinated they could magnify the damage of a physical attack. For example, a cascading outage would be aggravated if operators did not get the information to learn that it had started, or if protective devices were disabled.

Personnel Vulnerability

Workforce issues are critically important to maintaining a reliable supply of electricity, particularly in the event of a terrorist attack. Utility employees and contractors interact with the electric power system as managers, operators, line-crews, suppliers of materials and services, and users, among other roles. Although workers and managers in this industry have an outstanding record of reliable performance, even a few pernicious people in the wrong place are a potential source of vulnerability should they choose to disrupt the system.

A second issue is that, to a greater extent than in most other industries, the electricity workforce is aging, and many skilled workers and expert engineers will soon retire. As the current workforce retires, utilities may have increasing difficulty hiring sufficiently qualified replacements to keep the system operating effectively and reliably and to undertake all the upgrades that are needed, let alone cope with damage from terrorist attacks. This issue requires sustained and high-level attention by both the industry and federal agencies.

REDUCING RISKS

Reduce Vulnerability

The extent of the damage from an attack can be limited by a variety of means, including improving the robustness of the system to withstand normal failures; adding physical and cyber protections to key parts of the system; and designing it to degrade gracefully after catastrophic damage, leaving as many areas as possible still with power. Research and development can make particularly important contributions in these areas. **Table S.1** lists examples of changes that could be made starting now and others that could become options in the long term. Many of the changes discussed in this report could convert an attack that today could cause a blackout over a wide region of the country into one that would do less damage to the electric system and leave the system in a better position to accommodate the damage that does occur. Cascading failures could be limited, and many areas within a blacked-out region could maintain power because they could isolate themselves from the failing grid and maintain a balance of generation and demand within their borders.

TABLE S.1 Examples of Options for Minimizing Vulnerability

	Selected Options Currently Available	Selected Options That R&D Could Make Available
Physical vulnerability	Hardening of key substations and control centers Increased physical surveillance Addition of transmission towers that can prevent domino-like collapse	Improved intrusion sensors Development of strategies to provide greater system capacity Greater use of distributed generation and micro-grids
Cyber vulnerability	Elimination of all non-essential pathways to external systems Use of high-quality cyber security on all links	Improved cyber security for sensors, communication, and control systems Systems to monitor for and help avoid, operator error
Personnel vulnerability	Improved employee and contractor screening Improved training for attack response Improved planning and coordination with government (especially law enforcement)	Improved training simulators Expansion of support for educational programs in power engineering that have atrophied in large part because of very limited research investment
Increased system robustness and graceful degradation	A change in institutional arrangements and incentives to ensure adequate modernization of the transmission system Greater use of high-voltage power electronic technology Greater use of DC interconnects Expanded and more selective demand-side management and distribution automation	Lower-cost undergrounding Improved probabilistic vulnerability assessment Improved sensors, communication, real-time analysis, and system visualization Improved automatic control Improved capability for islanding and self-healing Improved energy storage
Accelerated restoration	Expanded planning for very large outages Designation of some utility employees as first responders.	Development and stockpiling of restoration transformers and other key equipment of long lead-time Improved assessment and planning tools.
Maintenance of critical services while grid power is disrupted	Use of robust systems such as light-emitting diode (LED) traffic lights with trickle charge batteries Co-location of generation with critical loads such as pumps for water supply Comprehensive contingency planning Avoidance of cross-dependencies (e.g., backup power for cell phone sites; gas rather than electric pumps on gas pipelines)	

Physical protection of critical facilities includes hardened enclosures for key transformers, improved electronic surveillance, and system tools that can identify physical and control system problems and potential incidents. Such measures may deter as well as blunt an attack.

Cyber security is best when interconnections with the outside world are eliminated. When interconnections are unavoidable, best practices for security must apply. Wireless communications within substations is a particular concern.

The risk of insider-assisted attacks can be reduced by strengthening background checks for new and existing employees and contractors. If subversive or disaffected workers can be identified, attackers will lose a major potential advantage. Training operators and other workers to recognize and react to attacks or other major disruptions will be helpful in limiting the extent of outages and further damage during a cascading failure. System simulators are likely to be very useful in this endeavor. In the long term, supporting engineering and other technical education will help to maintain the availability of the necessary skills in the workforce.

Even if terrorist attacks were not a concern, the transmission system should be modernized and upgraded to handle the increasing flow of power. A robust, modern system could ride out disturbances that would cause major problems to today's stressed system. The new operating standards being prepared by the electric industry and its reliability organizations under the Energy Policy Act of 2005 (EPAct) will help, but EPAct doesn't directly grant authority to order upgrades in the physical system. Industry, the Federal Energy Regulatory Commission (FERC), the Department of Energy (DOE), and state public utility commissions are aware of such needs, but building new transmission lines and other delivery enhancements is expensive and difficult. Upgrading sensors and controls can allow more power to flow on existing lines, which will help under some conditions. The terrorist threat suggests that additional upgrades may be important to reduce major outages. Current standards are met if no significant outage occurs following the failure of one major line or certain related double outages. Damage by terrorists could greatly exceed this level. A higher standard would be to maintain reliability when two major related failures occur, known as an N-2 event, which, in most cases, would entail additional costs. Improving the information flow to operators and the tools they can use to analyze and react to disturbances also would help prevent outages from cascading.

In the longer term, changes to the configuration of the power system could have dramatic impacts on its vulnerability. Among these, increasing generation within or close to major load centers, expanded use of distributed resources (cogeneration, micro-grids) with associated automatic control, and the successful development and deployment of storage technology would help limit cascading failures and leave islands of power within a blacked-out region.

Expedite Restoration

After an attack, an electric utility's main focus will be on restoring power to its customers. Many of the steps to be taken would be similar to those taken in response to a major natural disaster, such as a hurricane: that is, identify the damage, clean it up, repair equipment, and restore power. However, there are also important differences. Unlike hurricanes, terrorists may strike with no warning and selectively destroy the most important facilities, such as major substations. Some of the lost equipment may take months or even years to replace. Unless prior arrangements have been worked out, law enforcement officers might exclude utility workers from the crime scene while they investigate, delaying assessment of the damage and restoration activities. In addition, utility workers might be subjected to unexpected risks, such as chemical contamination.

Although detailed restoration plans cannot be formulated until specific damage is identified and the extent of an outage determined, advance planning can greatly speed the process of recovery. This is a well-established tenet in the industry. Utilities and transmission operating entities can- and do- make contingency plans. In preparing for a possible terrorist attack, they should set up an incident command system, establish good communications with government agencies, and reach agreements as to responsibilities and authority over various aspects of the restoration. Further work to address any specific issues that might arise in a terrorist incident is critical. Designating utility workers as first responders would improve their access to damaged substations and other facilities to assess the damage. Drills should be conducted for plausible scenarios of destruction to ensure that plans are adequate.

Key equipment, especially large power transformers, can be backed up with spares. The Edison Electric Institute (EEI) is developing the Spare Transformer Equipment Program (STEP), which will make spare transformers available in case of emergency. These transformers are very expensive, and not many spares are available. Transformers are also very large, heavy, and difficult to move. A major attack could quickly exhaust the inventory, and the world has limited manufacturing capacity. A promising solution is to develop, manufacture, and stockpile a family of universal recovery transformers that would be smaller and easier to move. These would be less efficient than those normally operated and so would only be for temporary use, but they could drastically reduce the delay before the electric system is back in full operation. Emergency backup policies also should be implemented for other key equipment such as large bushings and circuit breakers, which could take many weeks to replace.

Utility restoration workers need adequate food, water, fuel for vehicles, and other essentials that may not otherwise be available during an extended outage. Communication networks also may degrade or fail in an extended outage, and it is essential that utilities have backup systems available that can be operated without grid power.

In addition, utilities and transmission operators should ensure that sufficient generating plants have black-start capability. This is provided by units that can be started with no offsite power available, a likely situation in a widespread blackout.

Reduce Vulnerability of Critical Services in the Event of Outages

Society is becoming ever more dependent on electric power. While system owners and operators should do all that they reasonably can to ensure that their systems are able to withstand anticipated assaults from natural and human sources, there are practical limits to how much these highly distributed systems can be hardened. Even without the threat of terrorism, there is a risk of occasional power outages, some of which will have large spatial scale and may last for many hours or even days. Terrorism increases the probable extent and duration of such outages and could cause them to occur at particularly inconvenient or damaging moments.

Since the complete elimination of all possible modes of failure is simply not feasible, an important design objective (in addition to resilience and the ability to rapidly restore the system after a problem occurs) should be the ability to sustain critical social services while an outage persists. Thus, in addition to strengthening the grid, society should also focus on identifying critical services and developing strategies to keep them operating in the event of power outages-be they accidental or the result of terrorist attack.

Strategies for managing an extended outage will require detailed planning and preparation to ensure that critical facilities can continue to operate, either from the remaining grid or from emergency power systems. Metropolitan areas with high demand and high reliance on transmission to deliver power from distant generating stations should be of particular concern in this regard. Critical facilities (such as hospitals) often have emergency backup power generation capability, but some of these are only intended to operate for several days. An extended outage could easily exhaust the supply of fuel. Many critical service providers have no emergency power at all.

Although it is not reasonable to expect federal support for all local and regional planning efforts, the Department of Homeland Security (DHS) and/or the DOE should each initiate and fund several model demonstration assessments at the level of cities, counties, and states. These assessments should systematically examine a region's vulnerability to extended power outages and develop cost-effective strategies that can be adopted to reduce or, over time, eliminate such vulnerabilities. Building on the results of these model assessments, DHS should develop, test, and disseminate guidelines and tools to assist other cities, counties, states, and regions to conduct their own assessments and develop plans to reduce their vulnerabilities to extended power outages. To facilitate these activities, public policy and legal barriers to communication and collaborative planning will need to be addressed.

At a national level, DHS should perform, or assist other federal agencies to perform, additional systematic assessment of the vulnerability of national infrastructure, such as telecommunications and air traffic control, in the face of extended and widespread loss of electric power, and then develop and implement strategies to reduce or eliminate vulnerabilities. Part of this work should include an assessment of the available surge capacity for large mobile generation sources. Such an assessment should include an examination of the feasibility of utilizing alternative sources of temporary power generation to meet emergency generation requirements (as identified by state, territorial, and local governments, the private sector, and nongovernmental organizations) in the event of a large-scale power outage of long duration.

Government entities need to provide incentives (e.g., grants, fee-based awards, taxes, regulation) to support incremental costs associated with public and private sector risk prevention and mitigation efforts to reduce the societal impact of an extended grid outage. Such incentives could include incremental funding for those aspects of systems that provide a public good but no private benefit and the development and implementation of building codes or ordinances that require alternative or backup sources of electric power for key facilities.

THE IMPORTANCE OF INVESTMENT IN RESEARCH

There are many technologies and strategies that could be employed to make the power system more robust in the face of terrorist attack, make service restoration more timely after an attack, and continue the provision of critical services while the power is out. The best way to make needed changes affordable, and to develop new, even more effective and affordable approaches, is through research. Chapter 9 of this report discusses the current state of research for electric power, along with a set of recommendations for addressing research needs and developing related strategies.

The research that is needed to address the problems of terrorism is, for the most part, the same as the research that would address the broad problems faced by the transmission and distribution grid. The recovery transformer noted above is one of the few exceptions of terror-specific technologies that should be pursued. For example, the advanced computational system under development to improve control of flows on the grid also would be very useful in minimizing a cascading failure after a terrorist attack. The committee reached this conclusion in part from an informal questionnaire the committee developed and distributed to leading technical experts in the field. This questionnaire identified a variety of potential short- and long-term R&D needs for transmission and distribution. Respondents were asked to prioritize needs first for the industry as a whole and then strictly in terms of reducing vulnerability to terrorism. With a few exceptions, the research needs in the two cases were identical.

The committee is very concerned that the level of actual investment in power system research is currently much smaller than it should be as measured according to a variety of societal metrics. However, agreeing on institutional arrangements that can significantly increase the levels of nongovernmental research investment in this field has been a persistent problem. Chapter 9 discusses one possible strategy, but the committee was unable to reach a unanimous view on how best to resolve this problem.

The level of protection for and resiliency of the electric power grid against terrorist attacks needs to increase. However, the level of security that is economically rational for most infrastructure operators will be less than the level that is optimal from the perspective of the collective national interest. Therefore, the DHS should develop a coherent plan to address the incremental cost of upgrading and protecting critical infrastructure to that higher level.

In the specific context of electric power delivery, the Department of Homeland Security should:

- **Recommendation 1** Take the lead and work with the DOE and with relevant private parties to develop and stockpile a family of easily transported high-voltage recovery transformers and other key equipment. Although the expected benefits to the nation of such a program are difficult to quantify, they would certainly be many times its cost if the transformers are needed (see Chapters 3, 6, and 9).
- **Recommendation 2** Work to promote the adoption of many other technologies and organizational changes, identified in this report, that could reduce the vulnerability of the power delivery system and facilitate its more rapid restoration should an attack occur (see Chapters 6 and 7).
- **Recommendation 3** Work with the power industry to better clarify the role of power system operators after terrorist events through the development of memoranda of understanding and planned and rehearsed response programs that include designating appropriate power-system personnel as first responders (see Chapters 7 and 8).
- **Recommendation 4** Offer assistance to the Federal Energy Regulatory Commission, to state public service commissions, and to other public and private parties in finding ways to ensure that utilities and transmission operators have appropriate incentives to accelerate the process of upgrading power delivery and eliminating its most obvious vulnerabilities (see Chapter 6).
- **Recommendation 5** Work with the Department of Energy and the Office of Management and Budget to substantially increase the level of federal basic technology research investment in power delivery. The committee notes that (1) much of what is needed has the nature of a "public good" that the private sector will not develop on its own; (2) current levels of research investment are woefully inadequate; and (3) most of the system's vulnerabilities to terrorism are integrally linked to other more general problems and vulnerabilities of the system and cannot be resolved in isolation (see Chapter 9).
- **Recommendation 6** Take the lead in initiating planning at the state and local level to reduce the vulnerability of critical services in the event of disruption of conventional power supplies, and offer pilot and incremental

funding to implement these activities where appropriate (see Chapter 8).

* **Recommendation 7** Develop a national inventory of portable generation equipment that can be used to power critical loads during an extended outage. Explore public and private strategies for building and maintaining an adequate inventory of such equipment (see Chapter 8).

GLOBAL TRENDS 2030
ALTERNATIVE WORLDS

A publication of the National Intelligence Council

NIC 2012-001

2012

Highlights:

"The recurrence intervals of crippling solar geomagnetic storms, which are less than a century, now pose a substantial threat because of the world's dependence on electricity."

"Until 'cures' are implemented, solar super-storms will pose a large-scale threat to the world's social and economic fabric."

Found online at http://www.dni.gov/files/documents/GlobalTrends_2030.pdf

EXCERPTS

POTENTIAL BLACK SWANS THAT WOULD CAUSE THE GREATEST DISRUPTIVE IMPACT (PG. XI):

Severe Pandemic: No one can predict which pathogen will be the next to start spreading to humans, or when or where such a development will occur. An easily transmissible novel respiratory pathogen that kills or incapacitates more than one percent of its victims is among the most disruptive events possible. Such an outbreak could result in millions of people suffering and dying in every corner of the world in less than six months.

Much More Rapid Climate Change: Dramatic and unforeseen changes already are occurring at a faster rate than expected. Most scientists are not confident of being able to predict such events. Rapid changes in precipitation patterns—such as monsoons in India and the rest of Asia—could sharply disrupt that region's ability to feed its population.

Euro/EU Collapse: An unruly Greek exit from the euro zone could cause eight times the collateral damage as the Lehman Brothers bankruptcy, provoking a broader crisis regarding the EU's future.

A Democratic or Collapsed China: China is slated to pass the threshold of US $15,000 per capita purchasing power parity (PPP) in the next five years or so—a level that is often a trigger for democratization. Chinese "soft" power could be dramatically boosted, setting off a wave of democratic movements. Alternatively, many experts believe a democratic China could also become more nationalistic. An economically collapsed China would trigger political unrest and shock the global economy.

A Reformed Iran: A more liberal regime could come under growing public pressure to end the international sanctions and negotiate an end to Iran's isolation. An Iran that dropped its nuclear weapons aspirations and became focused on economic modernization would bolster the chances for a more stable Middle East.

Nuclear War or WMD/Cyber Attack: Nuclear powers such as Russia and Pakistan and potential aspirants such as Iran and North Korea see nuclear weapons as compensation for other political and security weaknesses, heightening the risk of their use. The chance of nonstate actors conducting a cyber attack—or using WMD—also is increasing.

Solar Geomagnetic Storms: Solar geomagnetic storms could knock out satellites, the electric grid, and many sensitive electronic devices. The recurrence intervals of crippling solar geomagnetic storms, which are less than a century, now pose a substantial threat because of the world's dependence on electricity.

US Disengagement: A collapse or sudden retreat of US power probably would result in an extended period of global anarchy; no leading power would be likely to replace the United States as guarantor of the international order.

NATURAL DISASTERS THAT MIGHT CAUSE GOVERNMENTS TO COLLAPSE (PG.49):

In October 2011, the National Intelligence Council (NI C) partnered with Oak Ridge National Laboratory (ORNL) to identify and investigate natural disaster scenarios that would pose a severe threat to the US and other major nations. Participants—which included subject-matter experts from universities in the US, Canada, and Europe in addition to NI C and ORNL officials—were asked to distinguish among various categories of natural disasters: extinction-level events; potentially fatal scenarios with medium recurrence intervals; and "ordinary" disasters with short recurrence intervals.

Scenarios in the extinction-level category are so rare that they were discounted. The impacts from these events—such as large volcanic eruptions or impacts of large asteroids or comets—are likely to be minimal because something else—such as major military defeat or economic collapse—is far more likely to bring down any great nation or civilization. At the other end, "ordinary" disasters—which typically cause high mortality and substantial human misery and therefore warrant major prevention and recovery efforts—do not present a major threat to the foundations of nations or human society.

Far more serious threats are those natural disasters that are both sufficiently severe to bring down nations and also sufficiently likely to occur. A short list of candidates fitting these criteria includes:

Staple-crop catastrophes, especially extreme and prolonged drought, crop plagues, and highly sulfurous long-duration but low-level volcanic eruptions. Although severe outbreaks of generalist pests (locusts and grasshoppers) are possible, many of the worst epidemics can be traced to the development of monocultures, which is increasingly the case in modern agriculture. (See page 35 where we talk about the potential for the spread of wheat rust to have a devastating effect because of the lesser biological diversity of wheat.) The "Laki" eruption in Iceland in 1783-84 only lasted eight months, but the "dry fog" that was produced by its sulfurous plumes resulted in a hemispheric temperature drop of 1.0-1.5 degree Centigrade and widespread crop failures.

Tsunamis in selected locations, especially Tokyo and the Atlantic Coast of the US. Tokyo—which is at a low elevation—is the largest global city at greatest risk. The largest tsunami that could hit the US East would be due to an earthquake in the Puerto Rico area. The travel time for the tsunami to the East Coast is only 1.5 hours. The probability of another massive earthquake occurring in Puerto Rico within this century is over 10 percent.

Erosion and depletion of soils. Modern agriculture is eroding soil at rates at least 10-to-20 times faster than soil forms. Worldwide soil erosion has caused farmers to abandon 430 million hectares of arable land since the Second World War, an area the size of India. Increases in oil prices and thus end of cheap fertilizers means that maintaining agricultural productivity without healthy soil will become increasingly expensive and difficult.

Solar geomagnetic storms that could knock out satellites, the electric grid, and many sensitive electronic devices. The recurrence intervals of crippling solar geo-magnetic storms are less than a century and now pose a threat because of the world's dependence on electricity. Until "cures" are implemented, solar super-storms will pose a large-scale threat to the world's social and economic fabric.

Electric Grid Vulnerability

Industry Responses Reveal Security Gaps

A report written by the staff of Congressman Edward J. Markey (D-MA) and Congressman Henry A. Waxman (D-CA)

2013

Highlights:

"More than a dozen utilities reported 'daily,' 'constant,' or 'frequent' attempted cyber-attacks ranging from phishing to malware infection to unfriendly probes. One utility reported that it was the target of approximately 10,000 attempted cyber-attacks each month."

"Most utilities have not taken concrete steps to reduce the vulnerability of the grid to geomagnetic storms and it is unclear whether the number of available spare transformers is adequate."

"Only twenty independently owned utilities, six municipally or cooperatively-owned utilities, and eight federal entities reported owning spare transformers."

Found online at
http://democrats.energycommerce.house.gov/sites/default/files/documents/Report-Electric-Grid-Vulnerability-2013-5-21.pdf

EXECUTIVE SUMMARY

The last few years have seen the threat of a crippling cyber-attack against the U.S. electric grid increase significantly. Secretary of Defense Leon Panetta identified a "cyber-attack perpetrated by nation states or extremist groups" as capable of being "as destructive as the terrorist attack on 9/11."[34] A five-year old National Academy of Sciences report declassified and released in November 2012 found that physical damage by terrorists to large transformers could disrupt power to large regions of the country and could take months to repair, and that "such an attack could be carried out by knowledgeable attackers with little risk of detection or interdiction."[35] On May 16, 2013, the Department of Homeland Security testified that in 2012, it had processed 68% more cyber-incidents involving Federal agencies, critical infrastructure, and other select industrial entities than in 2011.[36] It also recently warned industry of a heightened risk of cyber-attack, and reportedly noted increased cyber-activity that seemed to be based in the Middle East, including Iran.[37]

Current efforts to protect the nation's electric grid from cyber-attack are comprised of voluntary actions recommended by the North American Electric Reliability Corporation (NERC), an industry organization, combined with mandatory reliability standards that are developed through NERC's protracted, consensus-based process. Additionally, an electric utility or grid-related entity may take action on its own initiative.

In light of the increasing threat of cyber-attack, numerous security experts have called on Congress to provide a federal entity with the necessary authority to ensure that the grid is protected from potential cyber-attacks and geomagnetic storms. Despite these calls for action, Congress has not provided any governmental entity with that necessary authority. In 2010, bipartisan cyber-security legislation known as the GRID Act passed the House of Representatives by voice vote. If enacted, this legislation would have provided the Federal Energy Regulatory Commission (FERC) with the authority to require necessary actions to protect the grid. However, this legislation did not pass the Senate and has not been taken up again by the House since that time.

To inform congressional consideration of this issue, Representatives Edward J. Markey and Henry A. Waxman requested information in January 2013 from more than 150 investor-owned utilities (IOUs), municipally-owned utilities, rural electric cooperatives, and federal entities that own major pieces of the bulk power system. As of early May, more than 60% of the entities had responded (54 investor-owned utilities, 47 municipally-owned utilities and rural electric cooperatives, and 12 federal entities). This report is based upon those responses, and finds the following:

1. The electric grid is the target of numerous and daily cyber-attacks.

- More than a dozen utilities reported "daily," "constant," or "frequent" attempted cyber-attacks ranging from phishing to malware infection to unfriendly probes. One utility reported that it was the target of approximately 10,000 attempted cyber-attacks each month.

* More than one public power provider reported being under a "constant state of 'attack' from malware and entities seeking to gain access to internal systems."
* A Northeastern power provider said that it was "under constant cyber attack from cyber criminals including malware and the general threat from the Internet…"
* A Midwestern power provider said that it was "subject to ongoing malicious cyber and physical activity. For example, we see probes on our network to look for vulnerabilities in our systems and applications on a daily basis. Much of this activity is automated and dynamic in nature – able to adapt to what is discovered during its probing process."

2. Most utilities only comply with mandatory cyber-security standards, and have not implemented voluntary NERC recommendations.

* Almost all utilities cited compliance with mandatory NERC standards. Of those that responded to a question of how many voluntary cyber-security measures recommended by NERC had been implemented, most indicated that they had not implemented any of these measures.
* For example, NERC has established both mandatory standards and voluntary measures to protect against the computer worm known as Stuxnet. Of those that responded, 91% of IOUs, 83% of municipally- or cooperatively-owned utilities, and 80% of federal entities that own major pieces of the bulk power system reported compliance with the Stuxnet mandatory standards. By contrast, of those that responded to a separate question regarding compliance with voluntary Stuxnet measures, only 21% of IOUs, 44% of municipally- or cooperatively owned utilities, and 62.5% of federal entities reported compliance.

3. Most utilities have not taken concrete steps to reduce the vulnerability of the grid to geomagnetic storms and it is unclear whether the number of available spare transformers is adequate

* Only 12 of 36 (33%) responding IOUs, 5 of 25 (20%) responding municipally- or cooperatively-owned utilities, and 2 of 8 (25%) responding federal entities stated that they have taken specific mitigation measures to protect against or respond to geomagnetic storms.
* Most utilities do not own spare transformers. Only twenty IOUs, six municipally or cooperatively-owned utilities, and eight federal entities reported owning spare transformers. While other utilities reported participation in various mutual assistance agreements or industry equipment sharing programs, none knew how many other utilities would claim contractual access to the same equipment in the event of a large-scale outage.

Endnotes

[1] William J. Perry, James R. Schlesinger, et al., America's Strategic Posture: The Final Report of the Congressional Commission on the Strategic Posture of the United States, Washington D.C., 2009, p. xix.

[2] Ibid, pp. 90-92.

[3] R. L. Gardner, "Electromagnetic Terrorism. A Real Danger, "Proceedings of the XIth Symposium on Electromagnetic Compatibility", Wroclaw, Poland, June 1998.

[4] W. A. Radasky, M. A. Messier and M. W. Wik, "Intentional Electromagnetic Interference (EMI) – Test and Data Implications", Zurich EMC Symposium, February 2001.

[5] P. O. Leach and M. B. Alexander, "Electronic Systems Failures and Anomalies Attributed to Electromagnetic Interference", NASA Report 1374, National Aeronautics and Space Administration. Washington, CC 20546-0001, July 1995.

[6] *Ibid.*

[7] *Ibid.*

[8] Workshop on "Electromagnetic Terrorism and Adverse Effects of High Power Electromagnetic (HPE) Environments", Proceedings of the 13th International Zurich Symposium and Technical Exhibition on Electromagnetic Compatibility, February 16 - 18, 1999.

[9] AMEREM'96, Albuquerque, New Mexico, May 27-31, 1996.

[10] EUROEM '98, Tel Aviv, Israel, June 14-19, 1998, and EUROEM 2000, Edinburgh, Scotland, 30 May – 2 June 2000.

[11] International Radio Scientific Union (URSI) General Assembly, Toronto, 1999.

[12] E. Rosenberg, "New Face of Terrorism: Radio-Frequency Weapons", New York Times, 23 June 97.

[13] "City surrenders to £400m gangs", The Sunday Times, London, 2 June 1996.

[14] V. M. Loborev, "The Modern Research Problems", Plenary Lecture, AMEREM'96, Albuquerque, New Mexico, USA, May 1996.

[15] D. Sawyer, "20/20 Segment on Non-lethal Weapons", American Broadcasting Company (ABC), aired in February 1999.

[16] M. Bäckström, C. Frost and P. Ånäs, "Förstudie rörande vitala samhällssystems motståndsförmåga mot elektromagnetisk strålning med hög intensitet (HPM)", Användarrapport FOA-R--97-00538-612-SE, August 1997, ISSN 1104-9154. Abstract in English, English title: "Preliminary Study on the Resistance of Critical Societal Functions Against Intense Electromagnetic Radiation".

[17] I. W. Merritt, U. S. Army Space and Missile Defense Command, "Proliferation and Significance of Radio Frequency Weapons Technology", testimony before the Joint Economic Committee, United States Congress, February 25, 1998.

[18] International Radio Scientific Union (URSI) General Assembly, Toronto, 1999.

[19] 2009 Long-Term Reliability Assessment, 2009-2018. NERC. Princeton, NJ. 2009. http://www.nerc.com/files/2009_LTRA.pdf

[20] "Transforming America's Power Industry: The Investment Challenge 2010-2030," Edison Foundation report prepared by the Brattle Group, November 2008. http://www.edisonfoundation.net/reports.htm#transforming

[21] Data extracted from NERC's 2009 Long-Term Reliability Assessment data.

[22] "Critical Infrastructure Protection (CIP)" section of NERC's "Reliability Standards for the Bulk Electric Systemsin North America" http://www.nerc.com/files/Reliability_Standards_Complete_Set.pdf

[23] NERC's Critical Infrastructure Protection Committee website at: http://www.nerc.com/page.php?cid=1|9|117|139

[24] Makhosi, T., G. Coetzee, Generator Transformer Damage In Eskom Network, EPRI

Workshop on Transformers and Geomagnetic Currents, Washington DC, Sept 23, 2004.

[25] J. G. Kappenman, Chapter 16 – " Geomagnetic Disturbances and Impacts Upon Power System Operations", The Electric Power Engineering Handbook, 2nd Edition, edited by Leonard L. Grigsby, CRC Press/IEEE Press, pages 16-1 through16-22, published 2007.

[26] Kappenman, J. G. An overview of the impulsive geomagnetic field disturbances and power grid impacts associated with the violent Sun-Earth connection events of 29–31 October 2003 and a comparative evaluation with other contemporary storms, Space Weather, 3, S08C01, doi:10.1029/2004SW000128. 2005.

[27] Kappenman, J. G. Great Geomagnetic Storms and Extreme Impulsive Geomagnetic Field Disturbance Events – An Analysis of Observational Evidence including the Great Storm of May 1921, Advances in Space Research, Volume 38, Issue 2, 2006, Pages 188-199,

[28] Makhosi, T., G. Coetzee, Generator Transformer Damage In Eskom Network, EPRI

Workshop on Transformers and Geomagnetic Currents, Washington DC, Sept 23, 2004.

[29] Kappenman, J.G., Chapter 16 – " Geomagnetic Disturbances and Impacts Upon Power System Operations", The Electric Power Engineering Handbook, 2nd Edition, edited by Leonard L. Grigsby, CRC Press/IEEE Press, pages 16-1 through16-22, published 2007.

[30] Kappenman, J. G. An overview of the impulsive geomagnetic field disturbances and power grid impacts associated with the violent Sun-Earth connection events of 29–31 October 2003 and a comparative evaluation with other contemporary storms, Space Weather, 3, S08C01, doi:10.1029/2004SW000128. 2005.

[31] Kappenman, J. G. Great Geomagnetic Storms and Extreme Impulsive Geomagnetic Field Disturbance Events – An Analysis of Observational Evidence including the Great Storm of May 1921, Advances in Space Research, Volume 38, Issue 2, 2006, Pages 188-199,

[32] ORNL-6665: Electric Utility Industry Experience with Geomagnetic Disturbances"; 1991

[33] A few transmission lines operate with direct current (DC), which requires conversation from alternating current (AC) at one substation and then back again at the receiving substation. DC also is used to interconnect the four major regions in the United States and Canada because its use avoids the necessity of keeping systems synchronized.

[34] http://www.defense.gov/transcripts/transcript.aspx?transcriptid=5136

[35] http://www.nap.edu/catalog.php?record_id=12050#toc

[36] http://www.dhs.gov/news/2013/05/16/written-testimony-nppd-house-homeland-security-subcommitteecybersecurity-hearing

[37] http://articles.washingtonpost.com/2013-05-09/world/39139314_1_senior-u-s-oil-and-gas-companies-iran

Made in the USA
Middletown, DE
10 July 2015